Wild Food Plants
of the South Pacific
and Southeast Asia

Santol (*Sandoricum koetjape*)

STEVE W. CHADDE

Wild Food Plants of the South Pacific and Southeast Asia

Steve W. Chadde

AN ORCHARD INNOVATIONS BOOK
ISBN: 978-1951682354

Ver. 2.0 (4/18/2020)

CONTENTS

Edible Ferns. 1

Edible Herbs 7

Edible Palms 15

Edible Grasses. 27

Edible Tubers 33

Plants Eaten as Greens 45

Edible Fruits. 79

Edible Seeds 111

Poisonous Plants 127

Plants Used to Stupefy Fish 135

Index 141

Breadfruit, Kamansi (*Artocarpus altilis*)
see page 83

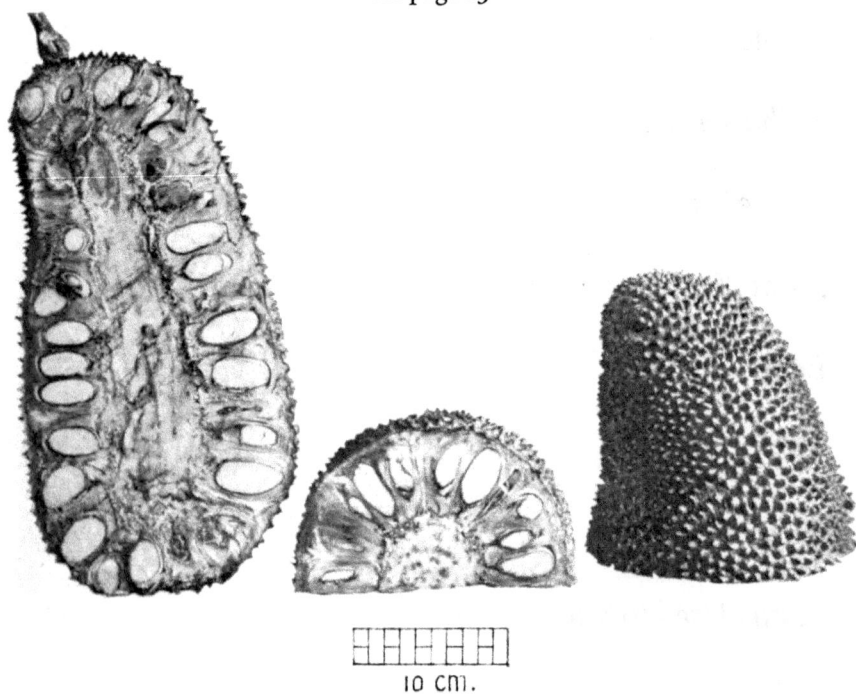

Jackfruit, Langka, Nanka (*Artocarpus integer*)
see page 84

INTRODUCTION

In 1943 the United States government published a manual "to aid the individual who becomes separated from his unit by illustrating and describing the edible and poisonous plants so that this individual can live off the land." At that time, in the midst of World War II, survival in the Tropics was a real issue for downed airmen, stranded sailors, and lost servicemen. A basic knowledge of some of the many plants safe to eat could mean the difference between life and death.

Today, as in the War years, the inhabitants of the Malayan and Polynesian regions use parts of a great many wild plants as food, sometimes to supplement and diversify their daily diet, and sometimes as famine foods in time of scarcity. The parts used include young shoots and leaves of various herbs, shrubs, and trees, various fruits, certain seeds, some flowers and flower buds, and the tubers or starchy bulblike roots of both cultivated and wild plants. Some of these plant parts have a high food value and some are rich in vitamins.

My hope is that this manual, updated with the scientific names currently in use, will serve as an introduction to the wild foods of the region, and provide the basic tools for identifying and preparing some of the interesting (and tasty!) food plants found in the South Pacific and Southeast Asia. I have retained much of the text of the original manual, however, as a precaution, only small amounts of any new wild food should be consumed initially until you are sure it is safe to eat. People living in the area often have a good knowledge of wild foods and should always be consulted for advice on proper methods of collection and preparation.

Scope

REGION COVERED. This manual covers all of Polynesia, Micronesia, and Melanesia, as well as the entire Malay Archipelago including the Malay Peninsula and the Philippines. For all practical purposes it also covers Indochina [the countries of Cambodia, Laos, Malaysia, Myanmar (Burma), Thailand (Siam), and Vietnam], and eastern India.

PLANTS. The more common plants that occur in reasonable abundance that may be used as food in times of emergency are included. Excluded are:

(1) Rare species.

(2) Plants that are familiar to residents of the temperate regions including maize or Indian corn, sorghum, rice, pineapple, cabbage, carrot, beet, garden bean, squash, cucumber, egg plant, sweet pepper, and other universally cultivated food plants.

(3) Familiar fruit trees such as the orange, lime, pomelo (one of the parents of the grapefruit), lemon, etc.

Reassurance and Warning

JUNGLE SNAKES. There is altogether too much fear of the Tropics, particularly on the part of those individuals without previous tropical experience. Thus the widespread fear of "the snake infested jungle" is an entirely imaginary picture. Poisonous snakes are essentially absent from Polynesia. In Malaysia, they are very rare and are seldom seen. The chances of being bitten by a poisonous snake in any part of the Malayan region are very much smaller than in any part of the United States where rattlesnakes and water moccasins occur.

POISONOUS PLANTS. There is no reason to fear the small number of poisonous plants in any part of Polynesia or Malaysia. The general rule is to avoid the following:

(1) Those with milky sap (except the numerous species of wild fig).

(2) All plants the taste of which is disagreeable.

CONTACT POISONS. In the Malayan and Polynesian region there are a few contact poisons corresponding to the poison ivy, poison sumac, and poison oak of temperate regions. However, they all belong to the same natural family of plants (Anacardiaceae). The poisonous principle and the treatment are the same as that indicated for persons coming in contact with poison ivy. In the Malay Peninsula, Sumatra, Java, and Borneo where most of them occur, they are collectively known as *rengas* and are all small to large trees. A few of the wild or semi-wild species of mango (but not the common mango) also have poisonous sap. These are sometimes cultivated or sometimes found in the forests, especially in the Malay Peninsula, Sumatra, Java, and Borneo. Normally an individual might be poisoned by these species when engaged in felling the trees. Their poisonous properties are well-known to the natives. Curiously, the fruits of all of these wild and semi-wild types of mango can be safely eaten, even though the sap is poisonous.

STINGING PLANTS. There are some types of plants, never very common, that have stinging hairs such as the tree nettles (*Laportea*) and the cowhage (*Mucuna*). The stinging hairs of the latter are merely mechanical irritants and are not poisonous.

OTHER PESTS. Keep in mind the fact that in all of Malaysia and Polynesia there is almost no danger from poisonous snakes, noxious insects, spiders, or poisonous plants. The forests and jungles of the entire region are a distinctly safe place under anything even approaching normal conditions. The malaria mosquito and the land leech are the major pests to avoid whenever possible. The land leech is found only in the high forests during the rainy season, or in areas where the rainfall is heavy in all months of the year.

Assistance and Advice of Natives

NATIVE USE OF PLANTS. In all parts of the region, the natives, in general, know both the wild and the cultivated plants which may be used as food. However, in certain sections, for example, Java, their use as food may be known, but quite unknown to the natives of other islands in Malaya, Micronesia, and Polynesia. The breadfruit, which is a basic food in many parts of Polynesia, is little used as food in most parts of Malaya where the species also occurs, simply because better foods are usually available there. A great many plants used by the natives of Java as food are quite unknown as food plants elsewhere.

ADVICE OF NATIVES. Whenever possible, try to get in touch with natives even though one may be able to talk with them only by means of signs. They can be most helpful in times when regular rations are not available. They usually know how these emergency food plants should be prepared, and those which may be poisonous if eaten raw. In some of the poisonous plants, the poisonous principle may be eliminated by proper cooking, or by other treatments, and the material then safely eaten.

Miscellaneous

PLANTS NEAR THE SEASHORE. The number and variety of plants on the atolls and low islands of Polynesia and Micronesia are usually small, whether the islands be small or large, inhabited or uninhabited. Naturally, a greater variety of food plants, many of them cultivated, are found on the inhabited islands. On most islands will be found on or near the seashore such plants as the pandan or screw pine, common purslane, seaside purslane, *Boerhaavia*, Polynesian arrowroot, and such shrubs and trees as *Ximenia*, *Morinda*, *Tournefortia*, *Pemphis*, *Thespesia*, and *Erythrina*, as well as various weedy herbs, such as *Alternanthera*, *Emilia*, *Amaranthus*, members of the Commelinaceae, and perhaps some other introduced weeds considered in this manual. Even on uninhabited islands is sometimes found the coconut palm and the breadfruit, where casual visitors have planted them. Generally, the vegetation on these low islands is very simple, with very few species as compared with that of the high islands such as Fiji, Samoa, and others, and with the large islands of the Malayan region.

GUIDE FOR EATING FRUITS. Keep in mind that those cultivated trees and shrubs growing in the settled areas, in and near towns, that bear attractive fleshy fruits, for the most part are actually planted for their fruits, and that generally their fruits may be eaten with perfect safety. In the wild, where monkeys occur, a safeguard to follow is to observe what the monkeys eat in the form of wild fruits. The feeding habits of birds is not such a safe guide. One should keep in mind that fruit maturity in the tropics is normally seasonal just as it is in temperate regions, and only occasionally, as with the coconut palm, are fruits produced throughout the year.

Acknowledgments

Much of this manual, published in 1943 as *Emergency Food Plants and Poisonous Plants of the Islands of the Pacific*, was prepared by Dr. Elmer D. Merrill (1876–1956), Administrator of Botanical Collections and Director of the Arnold Arboretum, Harvard University. Nearly all of the illustrations were prepared by Gordon W. Dillon (1912–1982). Their expertise is apparent on every page of the manual.

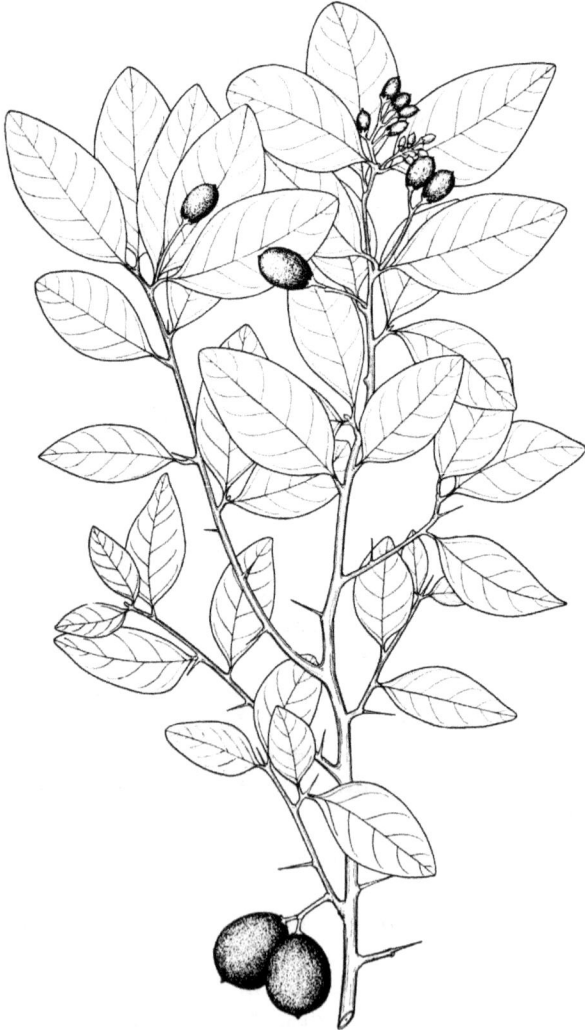

Tallow wood, Yellow plum, Sea lemon
(*Ximenia americana*), see page 108

Edible Ferns

"Fiddle head" of Paco (*Diplazium esculentum*)

Tree ferns reach large size in favorable environments

EDIBLE FERNS

Ferns in General

The number of different kinds of ferns in the Malayan-Polynesian region is very great, probably exceeding 1,500 different species. Some are small insignificant;, while others are relatively very large in size, including the characteristic tree ferns. Parts of certain species of ferns are regularly used as food by the natives and these parts are often offered for sale in native markets. While the food value of the edible parts of ferns is probably relatively low, these parts will help sustain life when other foods are not available.

In general, the parts most commonly used are the young unfolding leaves, commonly known as "fiddle heads;" these may be eaten either raw or cooked. Some of these fiddle heads are too tough, and others are bitter or otherwise bad tasting. But one point may be kept in mind that, so far as known, none of the ferns are actually poisonous when eaten. In some species the young tender leaves are cooked and eaten. In general only a few of the better known or useful ferns have definite plant names, but a common collective name for all ferns in the Malay Archipelago is *pako* or *paku*.

Tree ferns

Cyathea spp.   CYATHEACEAE

DESCRIPTION   Being chiefly forest ferns, tree ferns may sometimes be found in deserted clearings especially in more or less constantly wet regions. There are many different kinds and they are often abundant and are sometimes up to 25 feet high or even more.

PREPARATION   The young leaves as they commence to uncurl, the so-called "fiddle heads" are tender and may be eaten raw or cooked. The terminal tender bud or "cabbage" may also be eaten.

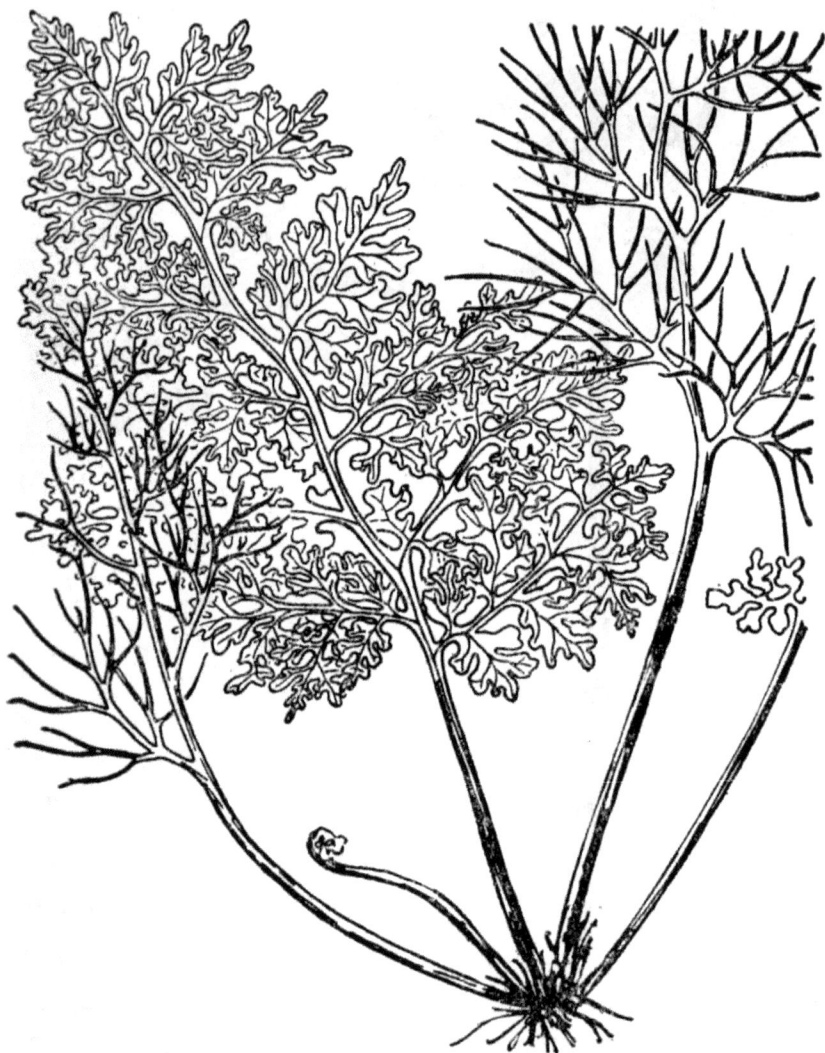

Swamp fern

Ceratopteris thalictroides (L.) Brongn.   PTERIDACEAE

DESCRIPTION	This fern, often occurring in great abundance, is found in very wet soil, old rice paddies, and swampy places, more or less submerged. It never occurs in salt or brackish swamps.
PREPARATION	The whole plant which is 1 to 12 feet high may be cooked and eaten as greens, or may be eaten uncooked. It is an excellent food.

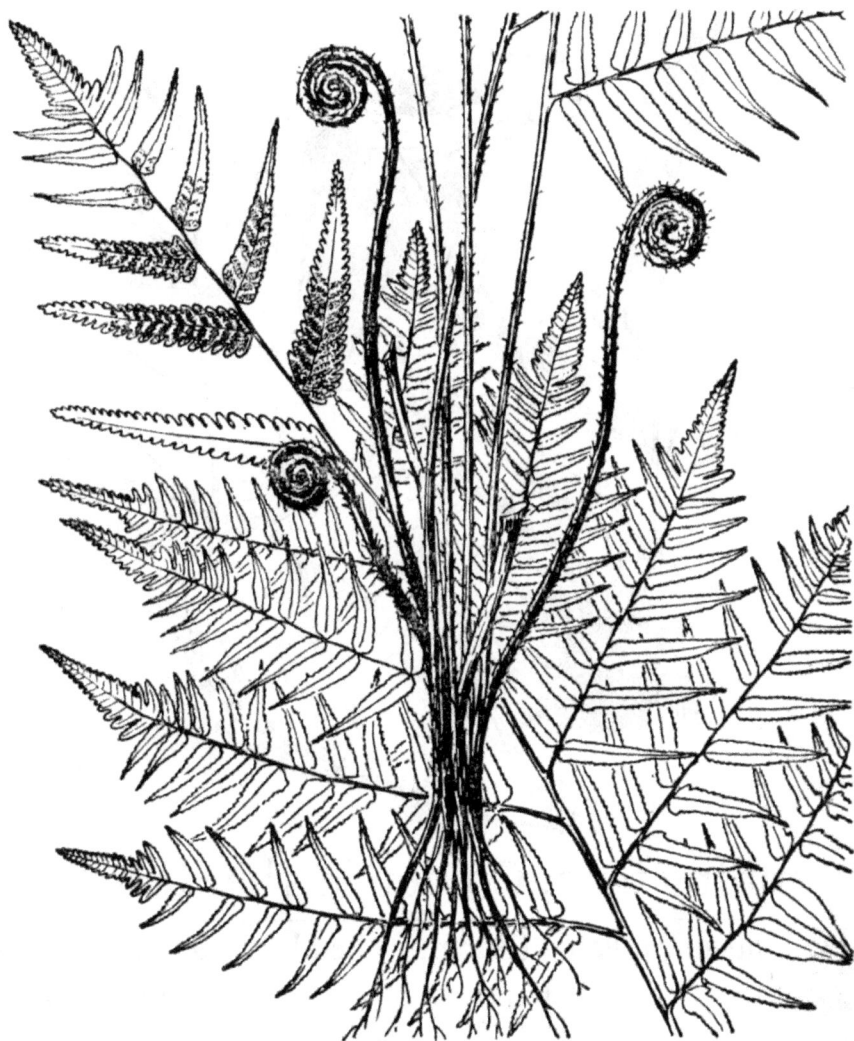

Paco

Diplazium esculentum (Retz.) Sw.   ATHYRIACEAE

DESCRIPTION   This fern often occurs in great abundance along swift-running streams, margins of rivers, and in some freshwater swamps. It is usually about 2 feet high.

PREPARATION   The young developing leaf stalks, or "fiddle heads," are an excellent food and may be eaten in quantity either raw or cooked.

Climbing fern, Golden leather fern

Stenochlaena palustris (Burm. f.) Bedd. (A) BLECHNACEAE
Acrostichum aureum L. (B)   PTERIDACEAE

DESCRIPTION   The *Stenochlaena* (A) is a climbing fern, occurring often in abundance near the inner margins of mangrove swamps, within the influence of salt or brackish water; other species occur in the inland forests. The *Acrostichum* (B) is a very coarse tufted fern, varying from 2 to 6 feet high, its mature leaves being very leathery. It grows only in brackish swamps and hence always near the seashore where it is commonly abundant.

PREPARATION   The tender young leaves of both these ferns may be cooked and eaten.

Edible Herbs

Taro (*Colocasia esculenta*)

EDIBLE HERBS

Araceae in general

These plants belonging in the calla lily family are found in the forests and in the open country, varying in size from small to very large herbs. None of the climbing ones should be used for food. Their vegetative parts are in general characterized by being supplied with myriads of minute needlelike stinging crystals of calcium oxalate that are intensely irritating when brought in contact with mucous membranes of the nose, mouth, and throat and, in some cases, even in contact with tender skin. These microscopic crystals (and they occur in our common Indian turnip or Jack-in-the-pulpit) cause the so-called acrid "taste" of these plants, but in spite of the very intense irritation they may cause, the plants are normally not actually poisonous.

In spite of the presence of these stinging crystals a considerable number of these plants are regularly eaten and several are widely cultivated for food, such as the taro (and the yautia in tropical America), and to a less degree the *Cyrtosperma* and *Alocasia*. In these cultivated forms, the underground part is usually greatly enlarged, forming a tuber very rich in starch. Thus to a very considerable degree these tubers take the place of the common potato in the Tropics where a starchy food is needed to help maintain a balanced diet.

The taro tuber in particular is a very excellent well-flavored vegetable. The taro leaves may be cooked and eaten, although the fresh leaves are abundantly supplied with the minute stinging crystals, which in the uncooked leaves are very irritating.

In general, when considering any of the numerous species of this family as food (other than the tubers of the taro), one should keep constantly in mind the usual presence of these microscopic stinging crystals of oxalate of lime and avoid putting any part of the raw plant into the mouth. The application of heat breaks down these stinging crystals so by thoroughly cooking the plant parts that are abundantly supplied with these very irritating needlelike crystals they may be safely eaten. However, in most cases, the first "taste" of the cooked aroid should be on the basis of a very small quantity, and if irritation results the material should be cooked for a longer period of time.

Taro

Colocasia esculenta (L.) Schott   ARACEAE

DESCRIPTION   This is one of the most commonly cultivated food plants in Polynesia, and also in the Malayan region, usually grown in wet lands. The many varieties are usually 1-1/2 feet high.

PREPARATION   The tubers are rich in starch and may be eaten in quantity, either boiled or roasted. They are an excellent substitute for the potato. The young leaves are commonly eaten as greens, but as they contain very many minute stinging crystal they must be thoroughly boiled before eating, as the application of heat destroys these irritating crystals.

Drop tongue

Schismatoglottis calyptrata (Roxb.) Zoll. & Moritzi   ARACEAE

DESCRIPTION   This low smooth herb grows as high as 2 or 3 feet, and its flowers are usually yellowish green, or the upper part is white. It occurs in moist shady places, especially in rocky soils, in forests, sometimes in thickets, and often near streams.

PREPARATION   All parts of the plant may be cooked and eaten.

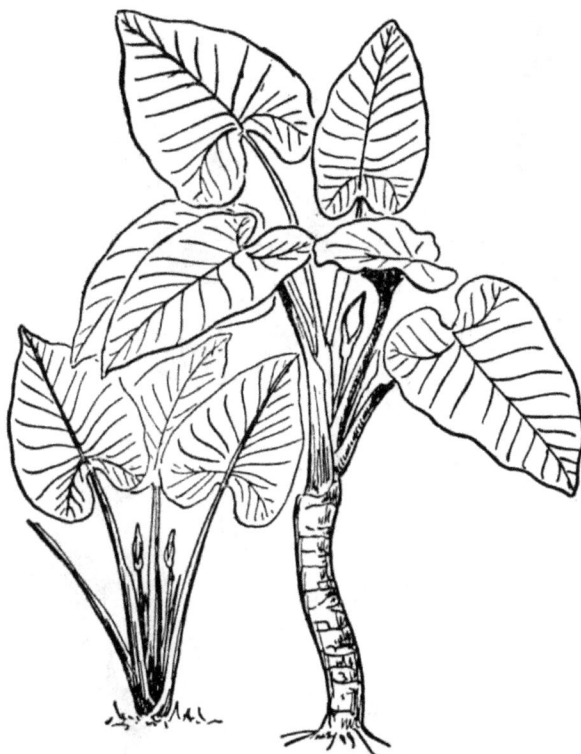

Elephant ear

Alocasia macrorrhiza (L.) G.Don   ARACEAE

DESCRIPTION	A very large plant growing in the forests and in open places, the elephant ear sometimes lacks a well-defined trunk, sometimes with a fairly tall trunk, and is often common and sometimes cultivated. It varies in height from 3 feet to as much as 12 feet.
PREPARATION	The juice is very acrid, due to the presence of thousands of tiny needlelike crystals of oxalate of lime. In contact with the nose and mouth they cause the most intense pain. In times of emergency the softer parts of the trunk, which contains considerable starch, may be cooked and eaten. Some varieties are much more irritating than others.
WARNING	Do not eat unless first cooked very thoroughly with two or three changes of water. Whenever possible, seek the advice of informed natives who know how to prepare the plant parts as food.

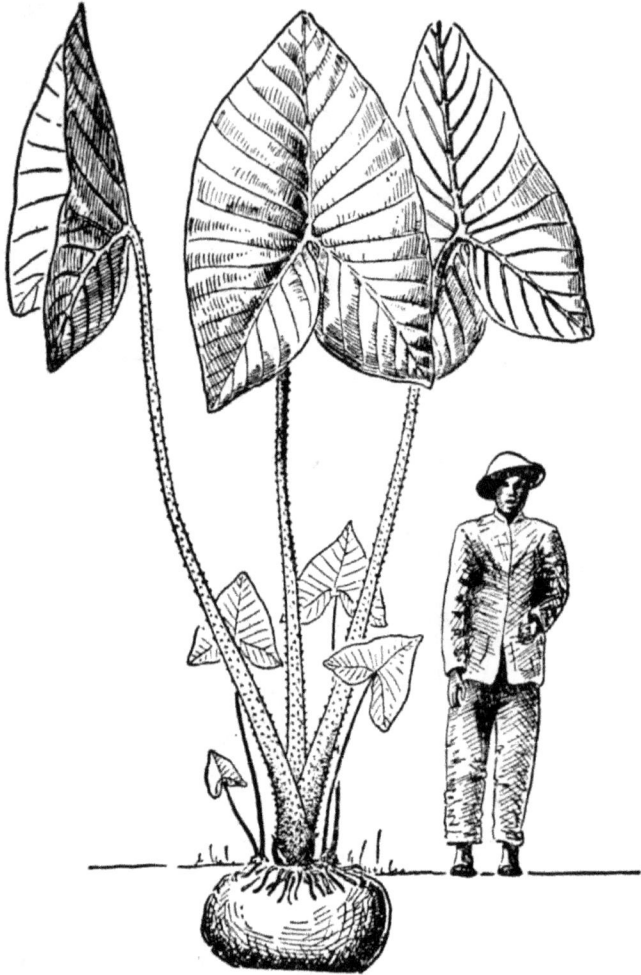

Giant swamp taro

Cyrtosperma merkusii (Hassk.) Schott   ARACEAE

SYNONYM	*Cyrtosperma chamissonis* (Schott) Merr.
DESCRIPTION	This is a very large plant growing only in fresh-water swamps or swampy places more or less in the open. The leaf stalks are more or less covered with short spines. Sometimes it is cultivated.
PREPARATION	The large underground part is rich in starch, but is to be eaten only when thoroughly cooked, either boiled or roasted.

Elephant foot yam

Amorphophallus paeoniifolius (Dennst.) Nicolson ARACEAE

SYNONYM	*Amorphophallus campanulatus* Decne.
DESCRIPTION	This plant, often common, has large flowers a foot or more in diameter appearing before the leaves. The flowers are purplish and mottled and have the odor of decaying meat. It is found in open places, near thickets, etc., and is sometimes cultivated. The characteristic leafy stem is usually about 3 or 4 feet high.
PREPARATION	The tender, young, rather rough and grayish mottled leaf stems may be eaten, but only after cookings The large tuber is rich in starch, but it contains innumerable minute stinging needlelike crystals which are intensely irritating to mucous membranes.
WARNING	The large tuber should never be eaten except after prolonged cooking. Long cooking breaks down the stinging crystals. If possible, consult informed natives before using this as food.

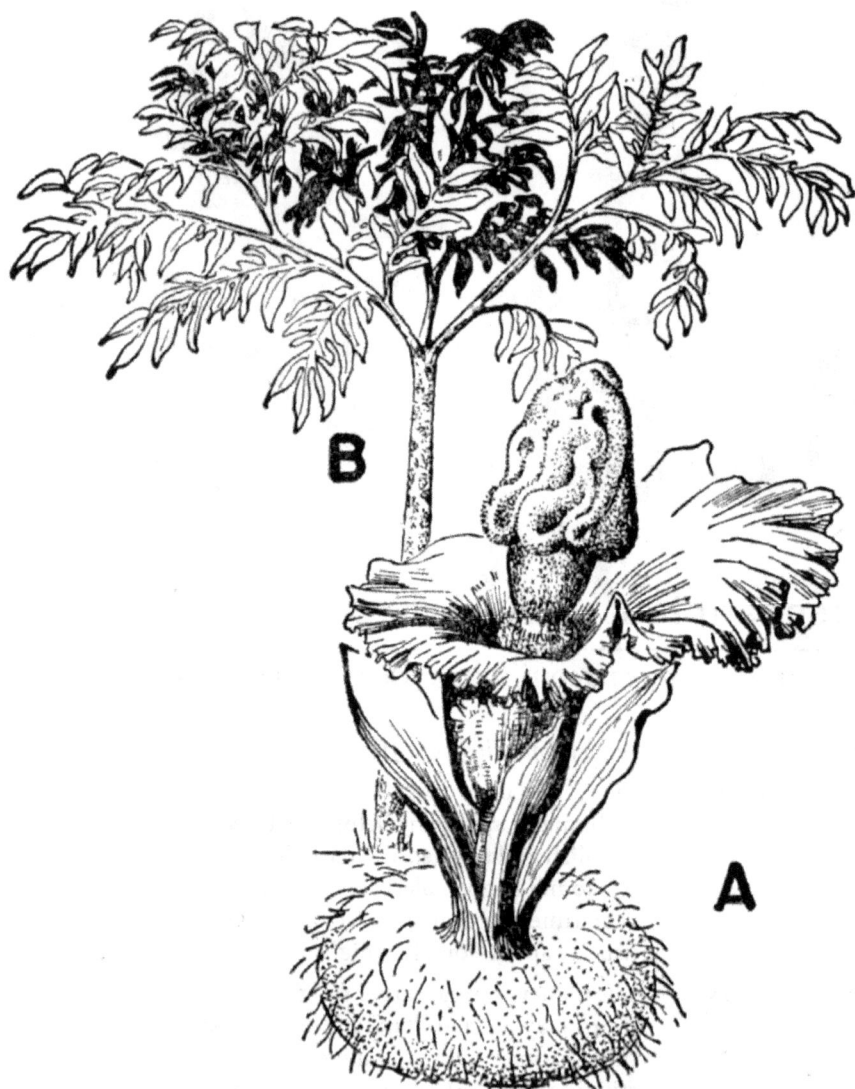

Amorphophallus paeoniifolius
(A) flowers, (B) leaves

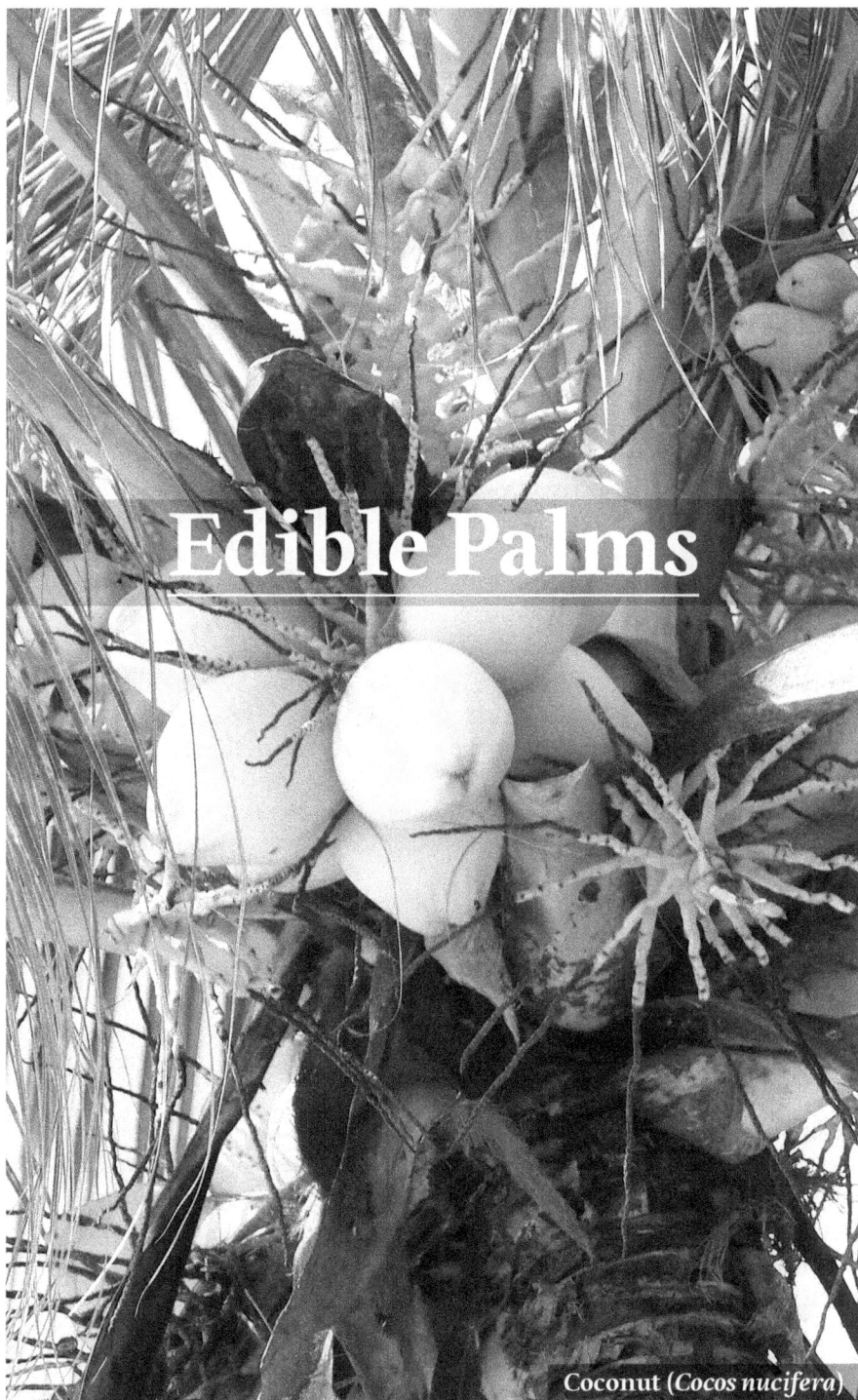

Edible Palms

Coconut (*Cocos nucifera*)

EDIBLE PALMS

Palms in general

There are a great many different palms In Malaysia and in Polynesia. They vary greatly in size and In habit. Some are very tall climbers, such as the rattan palms, others low and almost shrubby, and still others are gigantic in size. Some species grow along the seashore within the influence of the salt water, such as the nipa palm, some in open country, others in the secondary forests and thickets, and still others In the high forest.

Representatives of several genera (*Corypha, Arenga, Caryota, Metroxylon*) store up great quantities of starch In their trunks. This starch is entirely used up by the plant when it produces flowers and fruits, after which the plant dies. This starch is a valuable food, that from *Metroxylon* being the commercial sago. The starch from all of these palms is used for food. The palms are felled, split, the softer inner parts of the trunk crushed, and the starch washed out into troughs to settle. The water is then drawn off and the wet mass which dries is almost pure starch. The usual way of utilizing this starch for food is to make it into cakes which are then baked or roasted. The trunks of *Caryota, Metroxylon*, and *Arenga* are not large and can be manipulated rather easily; that of *Corypha* is gigantic, up to 3 feet in diameter, and the outside is very hard. In attempting to extract starch from any of these it is always best to enlist the services of natives. In any case select the trunks of palms that have not flowered, or, better, those that are just commencing to produce flowers.

In times of real emergency portions of the starch-bearing softer inner parts of these palm trunks may be cut into pieces which are then roasted or even boiled, after which the starch can be "chewed out" of the fibrous mass that forms the inner parts of the trunk. Its food value is high.

In general the terminal bud or "cabbage" of most palms is edible and may be eaten raw, boiled, or roasted. This palm "cabbage," except in those cases where it may be too bitter, is an excellent vegetable. This bud or "cabbage" is the actual growing tip of the trunk and is found deep in the terminal crown of leaf-stalk bases.

In the climbing rattan palms, which are particularly abundant in the high forests and of which there are many different species, the terminal bud or "cabbage" is edible; in many species the lower foot or so of the small trunk contains considerable amounts of starch. In cases of emergency these lower parts may be cut off, roasted over a fire, and the starch then "chewed out." The abundant small fruits of some of the species may be eaten, but the pulp is acid and scanty.

Very excellent, clear, tasteless or nearly tasteless drinking water may be obtained from the very long stems of the rattan palms. Cut the stems into about 6- to 8-foot lengths and hold these upright; the water will flow in a small stream from the lower end. In a short time the flow will stop and when this happens cut about a foot off the top end, and the flow will commence again. Repeat until all of the water is obtained.

The rattan palms are all high climbers, mostly very spiny on the leaf stalks and leaves and with long slender whip-like spiny appendages, the spines forming characteristic sharp claws. The very long slender stems are smooth and of the same diameter throughout. The stems vary from 10 or 15 feet to several hundred feet in length.

Except for the coconut palm and a very few others, the fruits of most of the Old World palm species are not edible. In fact, those of *Arenga* and *Caryota* are very dangerous as they are charged with myriads of minute needle-shaped stinging crystals that cause intense pain when in contact with nose or mouth or even the tender skin.

Warning: In testing palm fruits as to edibility, try only a very small quantity first. If immediate intense pain results, this means the presence of microscopic stinging crystals. Among the palms it is chiefly in the fruits of *Caryota* and *Arengra* that these intensely irritating stinging crystals occur, and one should never attempt to eat the fruits of these particular palms.

Sago palm

Metroxylon spp.

ARECACEAE

DESCRIPTION	This palm is found chiefly In fresh-water swamps and is one of the very few palms growing in such places. The tree is usually 25 to 30 feet high and smooth and very spiny forms occur. Fruit illustrated above.
PREPARATION	The trunk contains great quantities of starch which is the commercial sago and which is a basic food for the natives in many parts of Malaya. the starch-bearing softer inner parts of these palm trunks may be cut into pieces which are then roasted or even boiled, after which the starch can be "chewed out" of the fibrous mass that forms the inner parts of the trunk. Its food value is high.

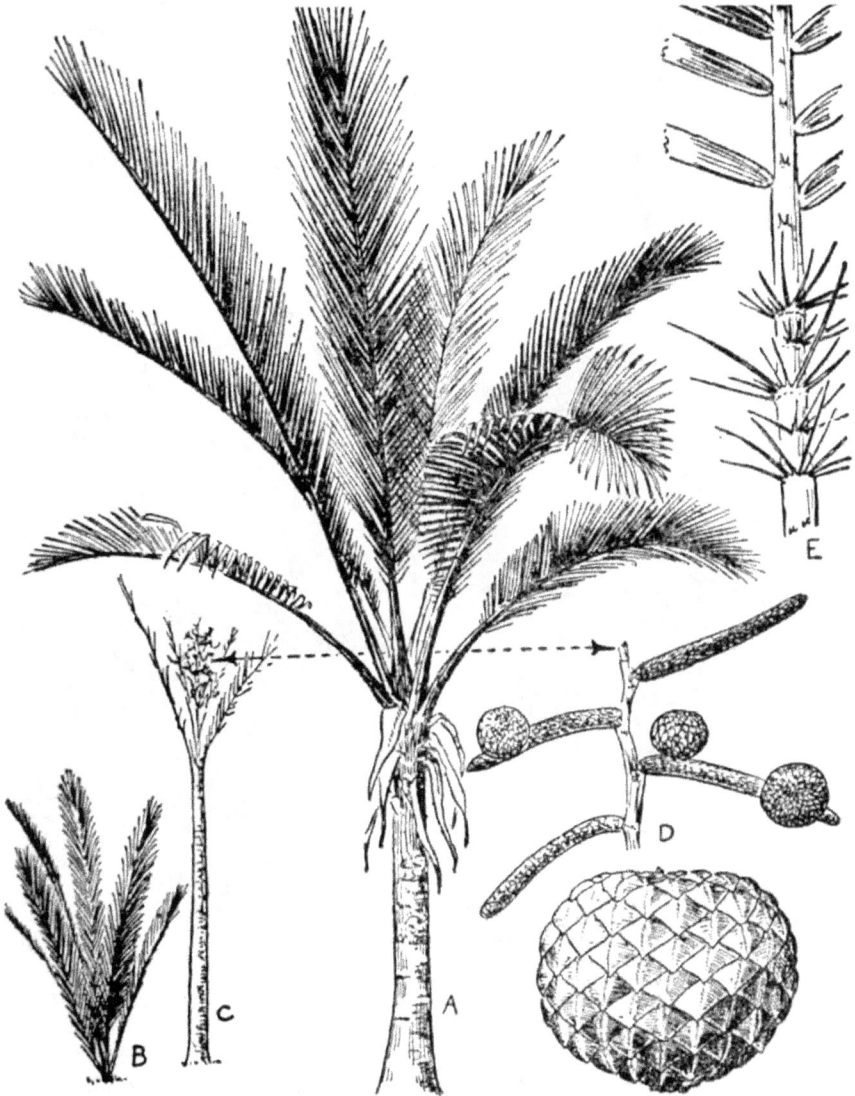

Sago palm (*Metroxylon* spp.)
(A) full-grown palm; (B) young palm;
(C) a palm past maturity in fruit, the starch in the trunk
all used up by plant; (D) flower and fruit bearing parts and
a mature fruit; (E) lower part of a leaf of the spiny form.

Salak

Salacca zalacca (Gaertn.) Voss   ARECACEAE

DESCRIPTION

This is a tufted, spiny, almost stemless palm which grows up to 15 feet high. It bears round, brown fruits which are covered with scales. Normally this palm is not found in forests but usually is planted.

PREPARATION

The yellowish white, sour-sweet, edible pulp surrounding two or three rather large hard seeds may be eaten raw. The immature fruits may also be cooked and eaten.

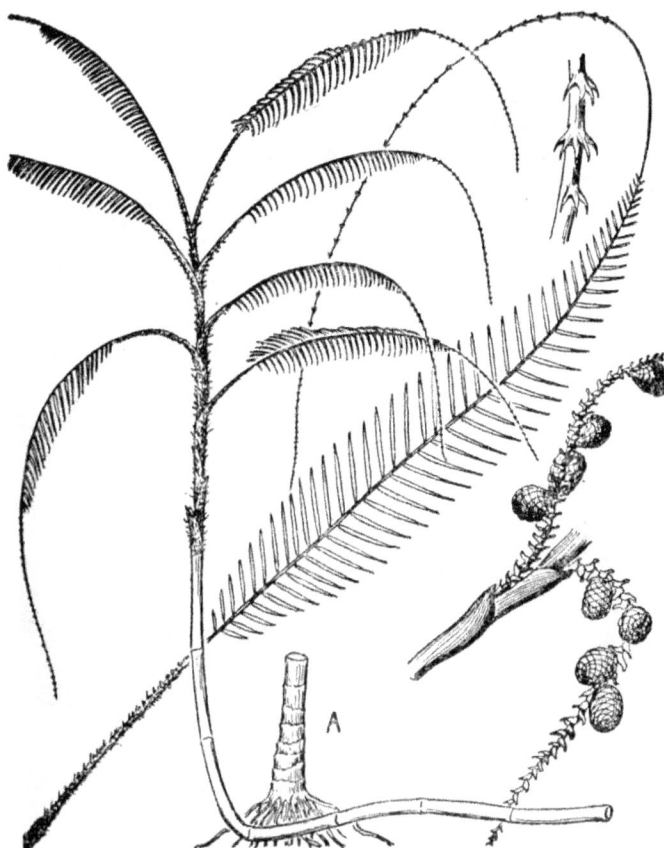

Rattan palm

Calamus spp. ARECACEAE

ARECACEAE

DESCRIPTION   There are many different kinds of rattan palm and they are found chiefly in the high forest. They are all climbing palms. The leaf stalks and growing parts are very spiny; the stems are smooth and vary in size from the thickness of a pencil to 2 inches in diameter. They are often several hundred feet long. The leaf tips are greatly extended and supplied with numerous very sharp, hard, claw-like teeth.

PREPARATION   The small growing point or "cabbage" of most species is edible. In many species the lower foot or two of the trunk (A) is slightly thickened and contains some starch; these lower parts may be roasted and the cooked starch "chewed out." The stews yield excellent drinking water.

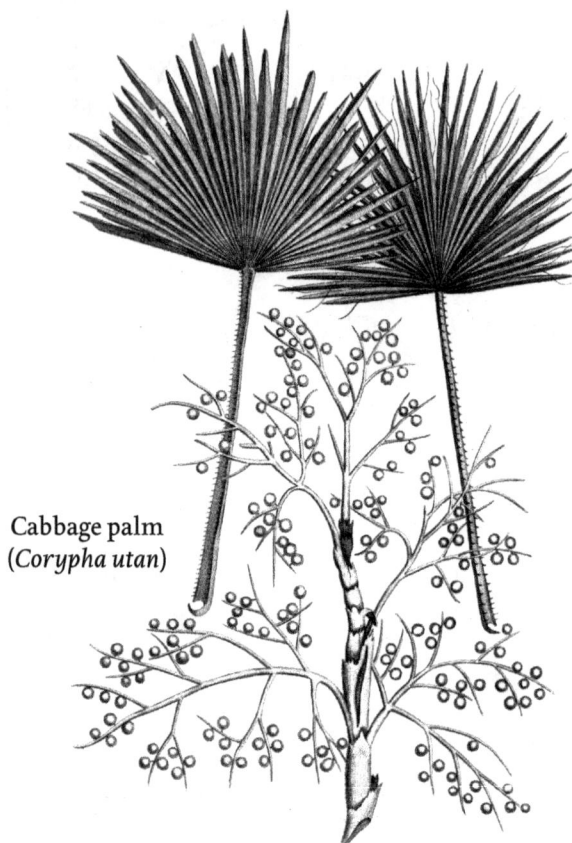

Cabbage palm
(*Corypha utan*)

Buri palm, Fishtail palm, Sugar palm

Species of *Corypha, Caryota, Arenga*   ARECACEAE

DESCRIPTION   All of these occur in open lands and in secondary forests;
the fishtail palm occurs also in the high forest. The buri
palm is recognized by its enormous size, often 50 feet
high, its great fan-shaped leaves, and their very stout
spiny leaf stalks; the fishtail palm by the shape of its
leaves; and the sugar palm by its very long feather-shaped
ascending leaves, the lower pants of the leaf stalks where
they join the stem with many very long black stiff hairs.

PREPARATION   All of these palms, like the sago palm, store up great
quantity of starch in the softer inner parts of their trunks
which may be used as food. The tender buds or "cabbage" of
all may be cooked and eaten.

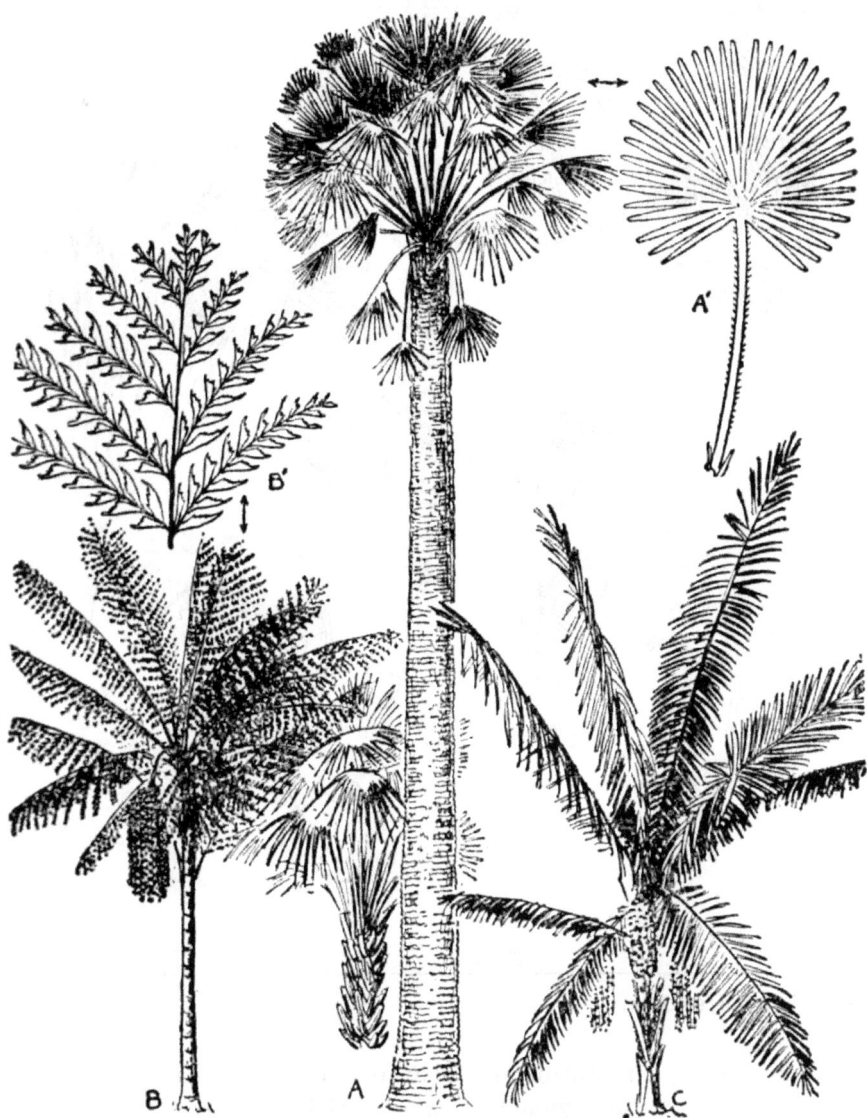

(A) Buri palm (*Corypha*), a young palm at the left;
(B) Fishtail palm (*Caryota*); (C) Sugar palm (*Arenga*)

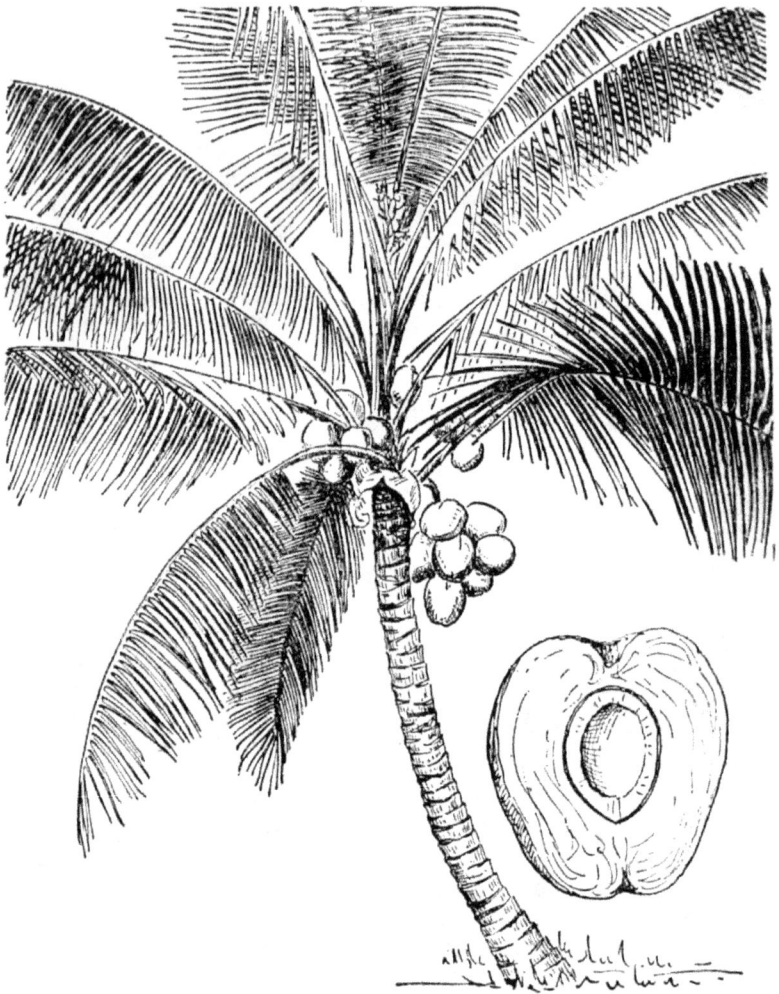

Coconut

Cocos nucifera L.

ARECACEAE

DESCRIPTION This plant is one of the most commonly cultivated palms throughout Polynesia and Malaya and needs little introduction.

PREPARATION The large terminal bud or "cabbage" is one of the very finest vegetables, and may be eaten in quantity, either raw or cooked. The nut yields the very best drinking water that is available anywhere, while the meat may be eaten in any stage of development.

Nipa palm

Nypa fruticans Wurmb ARECACEAE

DESCRIPTION This palm occurs only along tidal streams and back of the mangrove swamps where it is always within the influence of salt or brackish water. In favorable habitats it sometimes covers hundreds of acres. It is a stemless palm, the part corresponding to the trunk creeping in the mud and sending up several long leaves. The normal height is about 15 feet. The solitary, dark, brown, round heads of fruits are about 1 foot or more in diameter.

PREPARATION The large white seeds may be eaten when immature; in young stages they somewhat resemble the meat of the coconut. When fully mature the seed is very hard, and if eaten at all in this stage should be finely grated or crushed.

Nipa palm (*Nypa fruticans*)

Edible Grasses

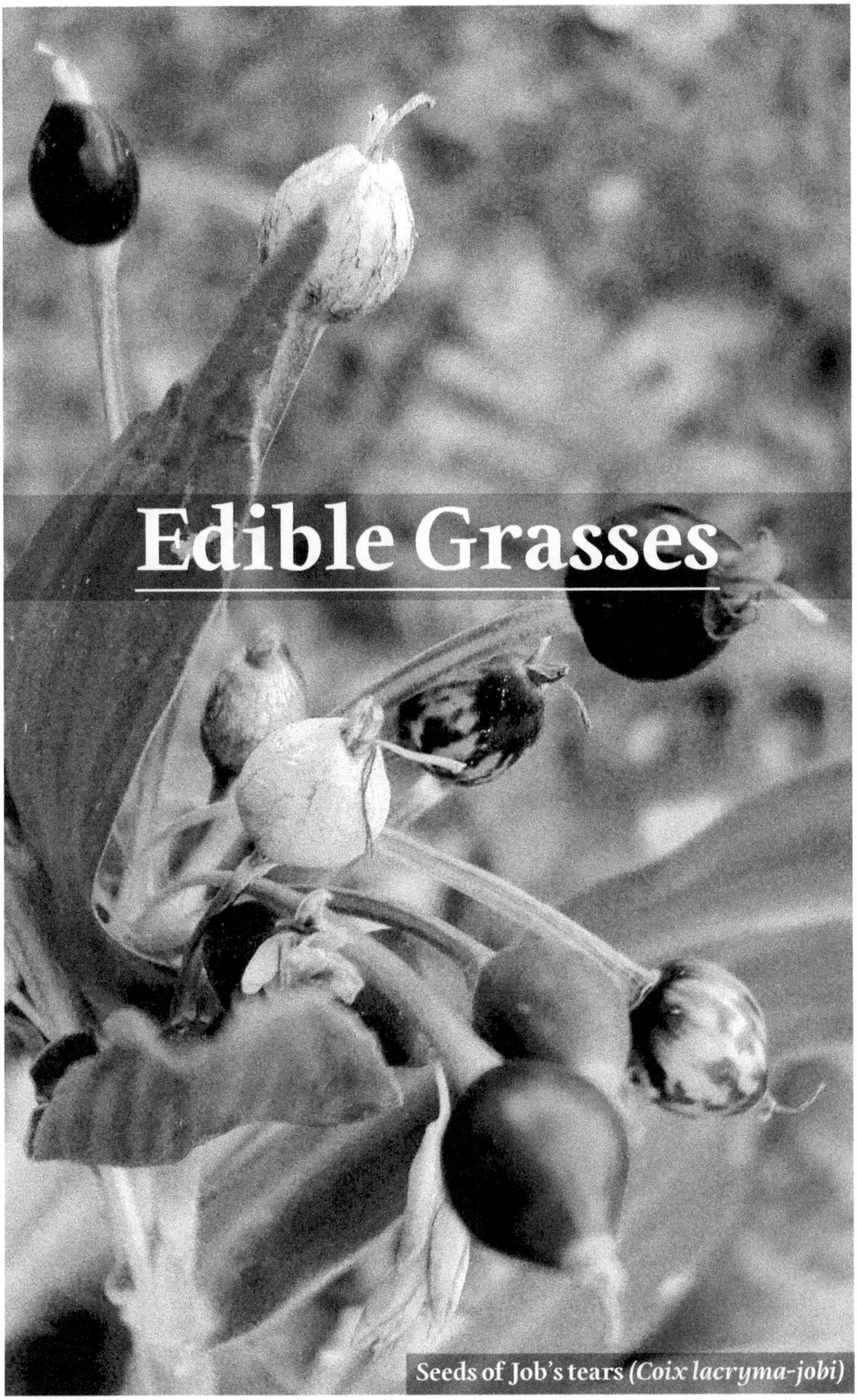

Seeds of Job's tears (*Coix lacryma-jobi*)

EDIBLE GRASSES

Grasses in General

To this family belong all of our cultivated cereals, such as rice, wheat, barley, oats, rye, millet, sorghum, maize or corn, etc. Rice,, millet, sorghum, maize, and several other cereals are extensively cultivated in the Tropics, but one does not find rye, wheat, oats, and other cereals so characteristic of the temperate regions. The bamboos are all grasses, and the young shoots of most of these (and there are many kinds in Malaya) may be cooked and eaten with safety.

The cultivated sugarcane is a grass. Its juice is rich in sugar, and thus has considerable food value. A wild species of sugarcane, a coarse, harsh-leafed grass 4 to 10 feet high, or even taller in rich soil, is very common and widely distributed in open valley lands. The flower-bearing parts are white, and make the species very conspicuous. It sometimes occupies large areas and scarcely needs a description. The hearts of the young shoots are frequently eaten raw or cooked, and are even sold in the markets of Java. The very young flowering parts, while still enclosed In the upper leaf-sheaths, may be cooked and eaten, while the roots may be peeled and eaten and taste somewhat sweet like the cultivated sugarcane.

Some of the wild grasses allied to millet, such as our common barnyard grass, have fairly large seeds, and these are produced in abundance; they may be gathered, the seeds rubbed out of the chaff, and either boiled or roasted. While the seeds of the wild grasses are much smaller than those of our cultivated cereals, nevertheless they are perfectly safe to eat, and are used by the natives in times of food shortages.

Job's tears

Coix lacryma-jobi L. POACEAE

DESCRIPTION	This is coarse grass, usually 2 to 3 feet high, often abundant in open places, never found in forests.
PREPARATION	The very hard, white, shining "fruit" contains one to several fairly large seeds which may be eaten raw, boiled, or roasted. There is one form with very thin-walled brownish "fruits," frequently cultivated for its seeds.

Palmgrass

Setaria palmifolia (J.Koenig) Stapf   POACEAE

DESCRIPTION	This is a coarse grass, 2 to 6 feet high, with broad, prominently nerved leaves and very numerous flowers. It is usually abundant in old clearings, partly shaded ravines, old plantations, and along forest borders.
PREPARATION	The hearts of the young shoots or stout plants (A) may be eaten raw or cooked, and these are often sold in the native markets of Java. The very numerous small seeds (the species being allied to Italian millet) may be gathered and boiled or roasted; these are used as a famine food in the Philippines and elsewhere.

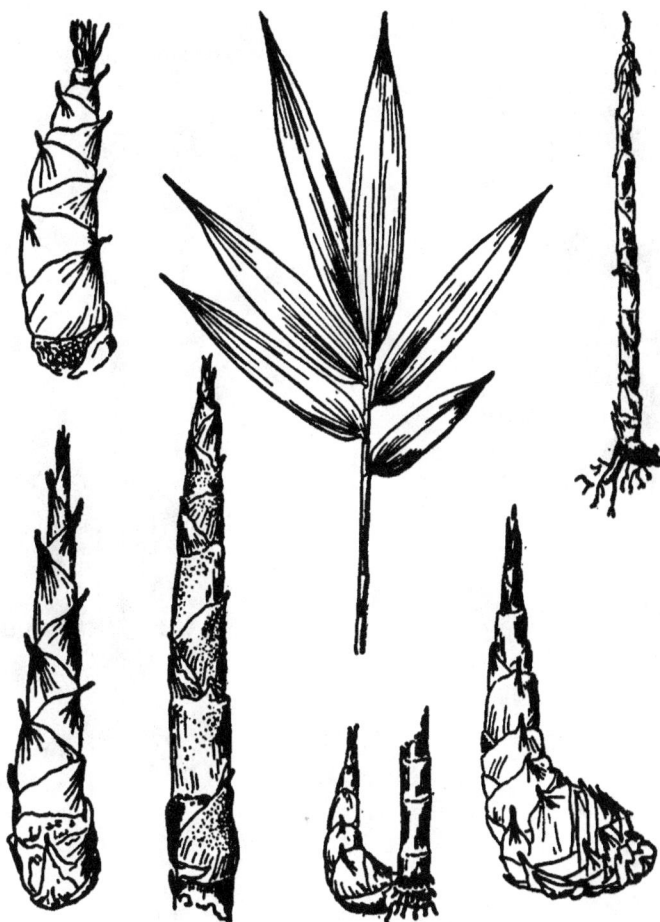

Bamboo

various species   POACEAE, SUBFAMILY BAMBUSOIDEAE

DESCRIPTION   There are many different kinds of bamboo in Malaysia and a few in Polynesia. They occur often in great abundance in the open country and in the jungles and forests.

PREPARATION   The young shoots appear from near the bases of the older stalks and their growth is very rapid. All of them may be cooked and eaten when young, although in a few species the shoots are too bitter to be palatable. The surrounding, often hairy sheaths, are removed and the more or less tender inner parts are cut into small pieces and boiled, or the whole shoot may be roasted.

Young bamboo shoot

Edible Tubers

Cassava (*Manihot esculenta*)

Sweet potato

Ipomoea batatas (L.) Lam.   CONVOLVULACEAE

DESCRIPTION   The sweet potato is widely cultivated throughout the Old World Tropics as a staple article of food. It may be identified by its pink flowers or the shape of its leaf.

PREPARATION   In addition to the edible tubers (these may be eaten raw or cooked), the young shoots and leaves make an excellent pot herb or substitute for spinach.

Cassava, Manioc, Tapioca

Manihot esculenta Crantz   EUPHORBIACEAE

DESCRIPTION	A plant widely cultivated in the Old World Tropics and is the commercial source of tapioca; it is a shrubby plant 3 to 5 feet high. The large roots are rich in starch.
PREPARATION	The two varieties, bitter cassava and sweet cassava cannot be distinguished by any characteristic other than by taste. Bitter cassava is poisonous when eaten raw. Cooking eliminates the poisonous principle (in this case hydro-cyanic acid), but with bitter cassava it is best to crush the root thoroughly and wash the starchy mass with several changes of water. Never eat bitter cassava raw, but only after it has been thoroughly cooked.

Greater yam

Dioscorea alata L.

DIOSCOREACEAE

DESCRIPTION | This is a twining vine, common in cultivation, and sometimes growing wild. The stems are ridged or with narrow wings. The yams vary enormously in shape and size, sometimes rather small, sometimes weighing up to 30 pounds.

PREPARATION | An excellent food boiled or roasted.

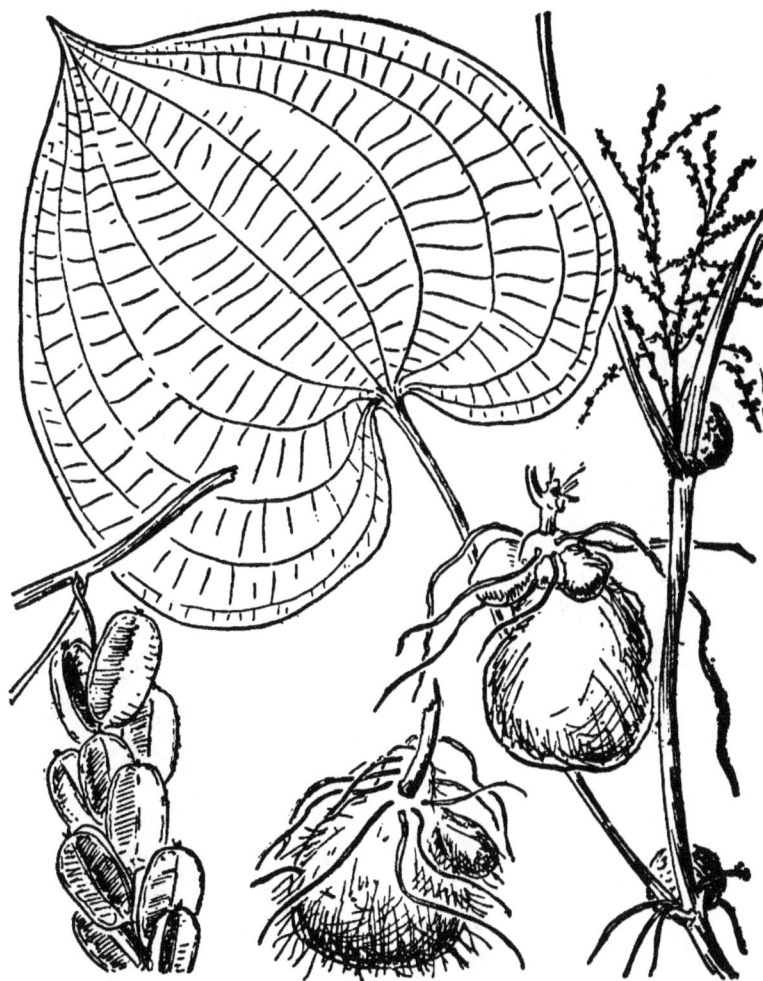

Bulb yam

Dioscorea bulbifera L.   DIOSCOREACEAE

DESCRIPTION   This twining vine has smooth stems. It grows in thickets, and is sometimes cultivated. Usually fairly large, round, rather hard bulbs are in the leaf axils.

PREPARATION   While the axillary bulbs and the yams may be eaten when properly prepared (see description for wild yam, *Dioscorea hispida*, p. 40), they should never be eaten unprepared, as they are definitely poisonous. Seek the advice of informed natives if possible as to how the tubers should be treated.

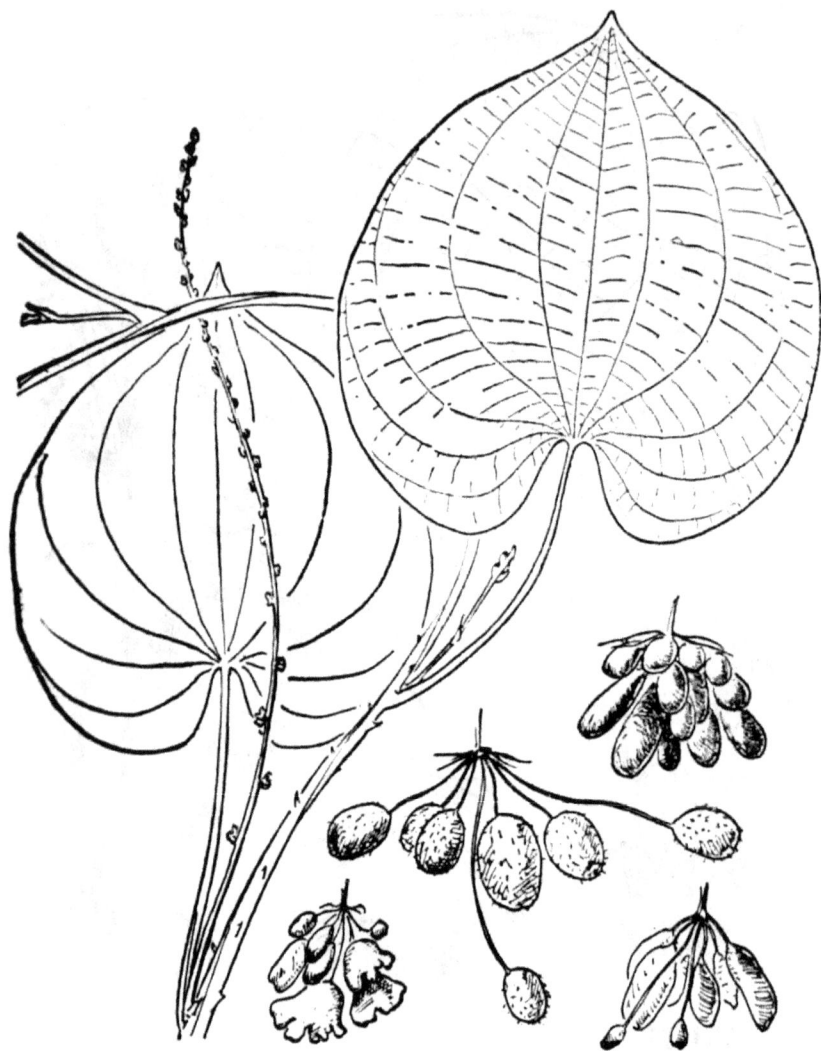

Goa yam

Dioscorea esculenta (Lour.) Burkill DIOSCOREACEAE

DESCRIPTION This spiny and twining vine is cultivated, also often found wild in thickets. The yams vary in shape but are usually not very large.

PREPARATION They are distinctly well flavored, and may be eaten boiled or roasted. Like those of the greater yam they need no special treatment, as they are never poisonous.

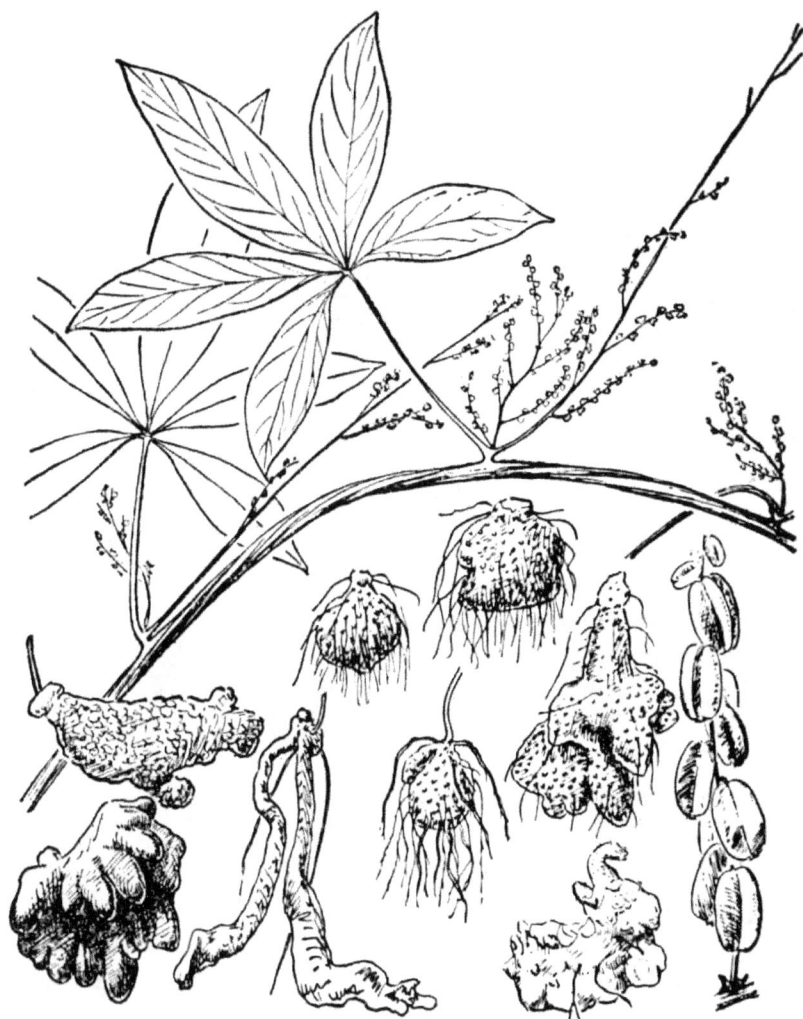

Buck yam

Dioscorea pentaphylla L. DIOSCOREACEAE

DESCRIPTION

This is a climbing, twining vine, the leaves usually with five parts;, the stems smooth or with short scattered spines. It is sometimes cultivated, but more commonly found wild in thickets. Sometimes there are small bulbs in the leaf axils. The yams vary in shape and are usually not very large.

PREPARATION

They may be eaten boiled or roasted.

Wild yam

Dioscorea hispida Dennst.　　　　　　　　**DIOSCOREACEAE**

DESCRIPTION　　This is a climbing, rather woody, spiny vine; the leaves have three parts. It usually grows wild in thickets, and is rarely cultivated. The yams vary considerably in shape and size.

PREPARATION　　The yams should be cut into very thin slices, coated with ashes if possible, and then soaked in streams or in salt water for 3 or 4 days, after which they should he dried in the sun. Alter prolonged treatment they may he cooked and eaten, hut great caution is necessary.

WARNING　　These yams are definitely poisonous and should not be used for food unless property prepared; seek the advice of natives whenever possible.

Arrowroot

Maranta arundinacea L.   MARANTACEAE

DESCRIPTION	This is an erect, smooth, branched herb, 1 to 3 feet high, with small white flowers. This is the commercial arrowroot, and is found only in cultivation.
PREPARATION	The thickened scaly roots may be cooked and eaten, or they may be crushed, the abundant starch washed out, and used as food.

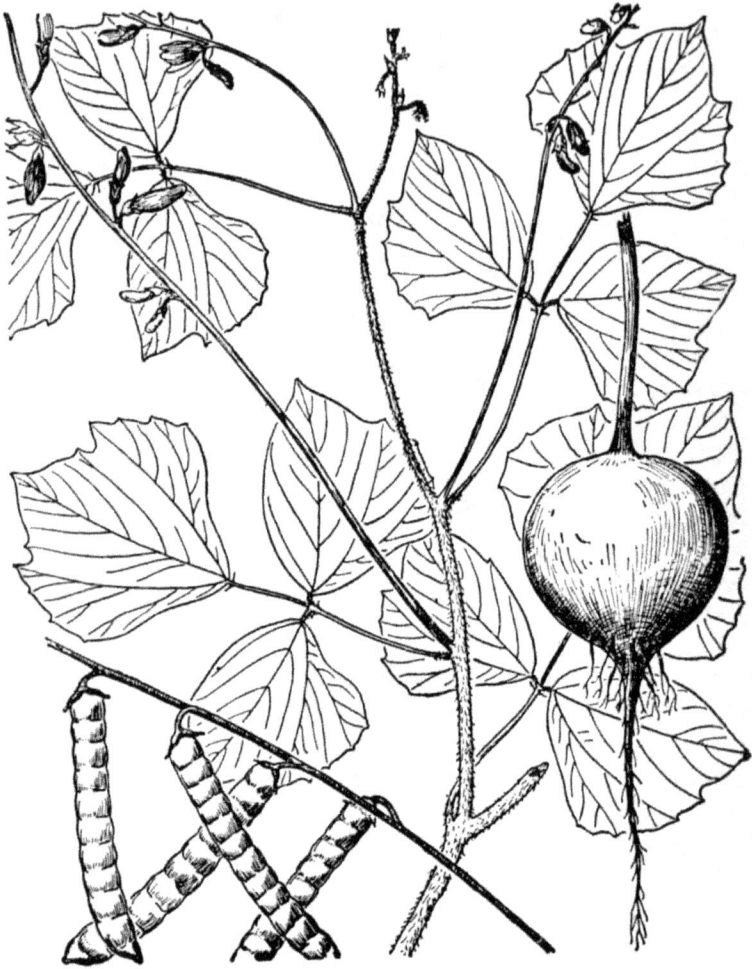

Yam bean

Pachyrhizus erosus (L.) Urb.   **FABACEAE**

DESCRIPTION	This vine has blue flowers. It is often common in thickets and hedgerows, and is sometimes planted.
PREPARATION	The turnip-shaped root is very refreshing, the flesh is crisp and pleasant to the taste; it is always eaten raw, never cooked. The very young pods may be cooked and eaten like string beans.
WARNING	The mature seeds in brown pods should never be eaten as these are poisonous.

Polynesian arrowroot

Tacca leontopetaloides (L.) Kuntze   DIOSCOREACEAE

DESCRIPTION	This is a plant that grows 2 to 5 feet high having stems that are distinctly grooved. Usually the tubers are found in the loose soil some distance from the base of the plant and from one to several to a plant. It is sometime cultivated, but as a wild plant is most often found in loose sandy soil not far from the seashore.
PREPARATION	The hard, usually round, and potato-like tubers are rich in starch and may be boiled or roasted and eaten, or better, crushed or grated and then boiled.
WARNING	The tubers should never be eaten raw as they are said to be poisonous until after being crushed, washed, and cooked.

Water chestnut

Eleocharis dulcis (Burm.f.) Trin. ex Hensch. CYPERACEAE

DESCRIPTION
This is a coarse, more or less tufted plant, 2 to 4 feet high, growing only in fresh water swamps in open places.

PREPARATION
The nearly round, hard tubers are produced underground and are excellent to eat, boiled or roasted. This is the wild form of the so-called *ma-tai* of the Chinese and the tubers in normal times are extensively imported into the United States by them and are served in Chinese restaurants.

Plants Eaten as Greens

Ceylon spinach (*Basella alba*)

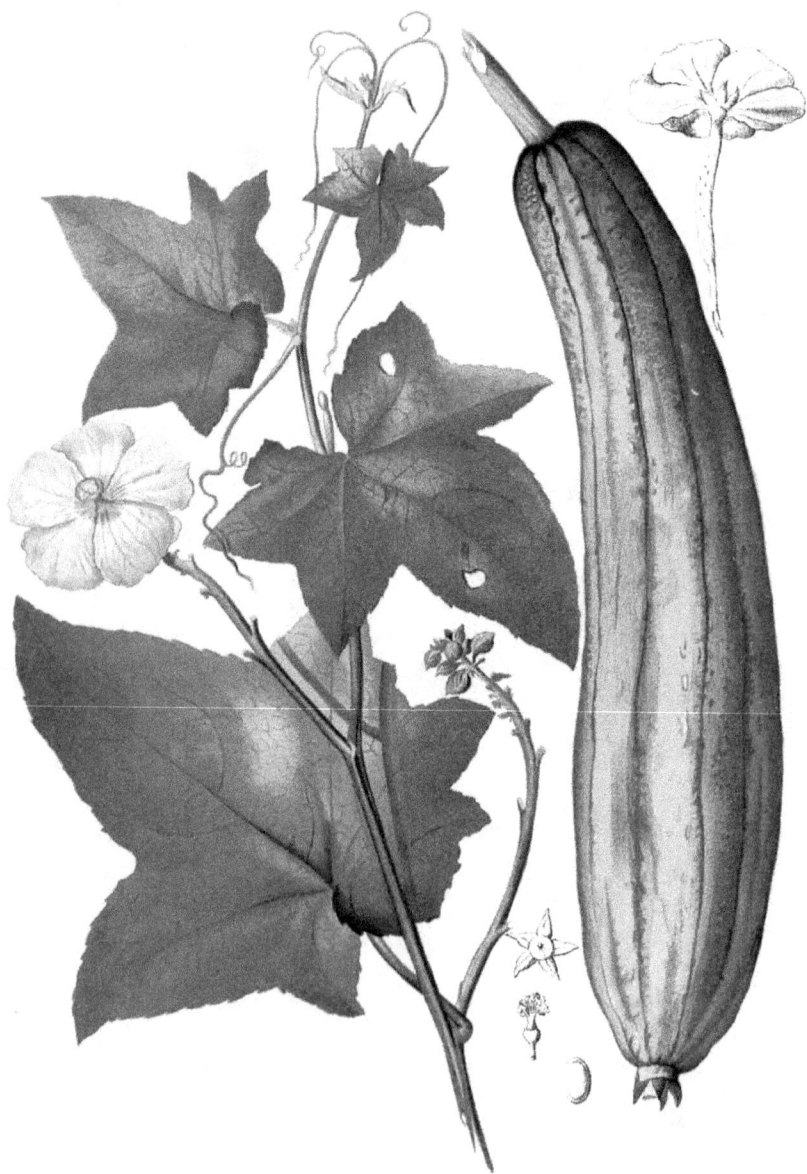

Luffa (*Luffa cylindrica*)

Luffa

Luffa cylindrica (L.) M.Roem.   CUCURBITACEAE
Luffa acutangula (L.) Roxb.

DESCRIPTION   These vines are cultivated, and also often grow wild. The flowers are yellow. *Luffa acutangula* (A) has sharply angled fruits; *Luffa cylindrica* (B) has smooth fruits suggesting a smooth cucumber. The fruits of the wild form, occurring in thickets especially near the sea, are smaller than those of the cultivated forms.

PREPARATION   The young green fruits (not more than half ripe) may be cooked and eaten; at this stage they make an excellent vegetable; the tender shoots, flowers, and young leaves may also be cooked and eaten. The mature fruits are too tough to eat.

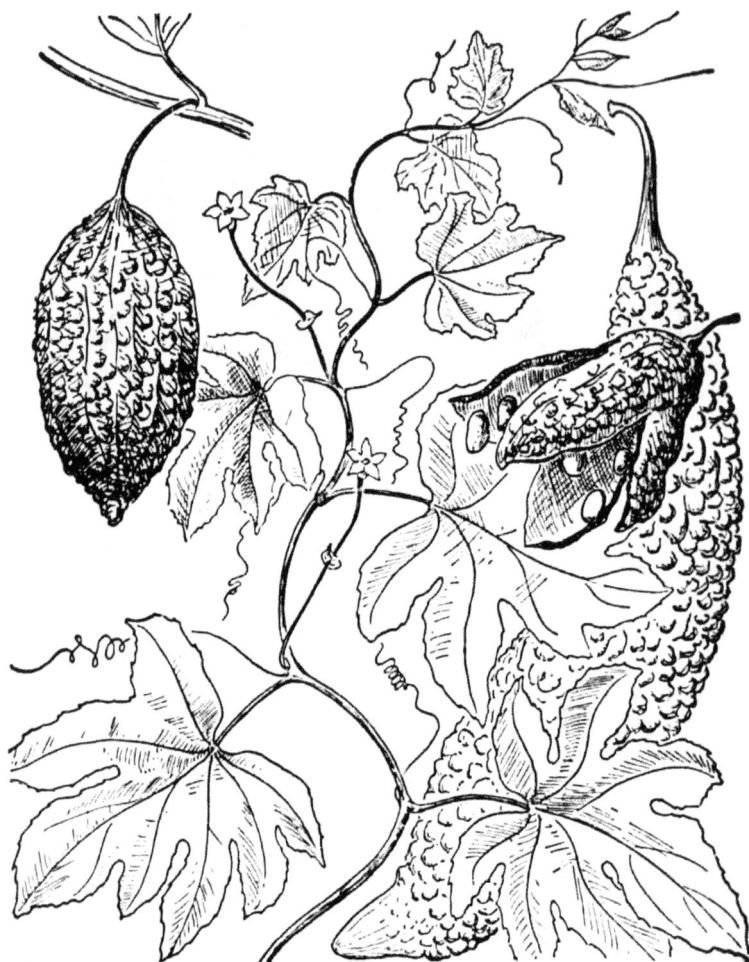

Balsam vine, Ampalaya

Momordica charantia L.   CUCURBITACEAE

DESCRIPTION   This is a slender vine with small yellow flowers. The rough
               fruits, variable in shape, are usually yellow, the pulp
               reddish. This plant is found both in cultivation and wild;
               the fruits of the wild form are always smaller than are
               those of the cultivated ones, which may be 6 inches long or
               longer.

PREPARATION   The young leaves and shoots may be eaten as greens (better
               mixed with other plant material, as they are rather bitter),
               while the fruits may be cooked and eaten.

Dayflower

Species of *Commelina* and *Cyanotis*   COMMELINACEAE

DESCRIPTION   Above are two common and widely distributed representatives of *Cyanotis* (A, D) and two of *Commelina* (B) (C). These are somewhat fleshy, trailing or ascending herbs, with blue flowers. They occur in open places, waste and cultivated lands and meadows, and all are common.

PREPARATION   The plants may be eaten raw, steamed, or boiled.

Amischotolype

Amischotolype mollissima (Blume) Hassk. COMMELINACEAE

SYNONYM	*Forrestia marginata* (Blume) Hassk.
DESCRIPTION	This erect plant grows to about 2 feet in height. The stems are smooth or hairy with dense heads of small violet or purple flowers in the leaf axils.
PREPARATION	The tender shoots may be cooked and eaten, these parts of the plant being even sold in native markets in Malaya.

Amaranth

Three species of *Amaranthus*   **AMARANTHACEAE**

DESCRIPTION   Various species of *Amaranthus* occur (often in great abundance) throughout the Malaysian and Polynesian regions, particularly in open places (especially B), waste lands about settlements (especially C), and more or less cultivated (A). Some (A and B) are often 3 feet high; others (C) are usually not more than 1 foot high. In the cultivated forms the leaves are often variegated, dull purple to even red.

PREPARATION   The young shoots and leaves of all kinds of Amaranthus make excellent greens when cooked.

Cock's comb

Celosia argentea L.

AMARANTHACEAE

DESCRIPTION (A) is a wild form of the common garden cock's comb and is often abundant in meadows, old clearings, waste places, but always in the open, never in forests. It is about 2 feet high and the flower bearing parts are shining white to pink.

PREPARATION The young shoots and leaves are boiled and eaten as greens. The garden forms of the common cock's comb (B) and (C) may also be so used, the floral parts being red, purple, or yellow, but these forms are usually not found wild.

Dwarf copperleaf

Alternanthera sessilis (L.) R.Br. ex DC. AMARANTHACEAE

DESCRIPTION	This is a common, widely distributed, weedy plant, more or less ascending. It is found in waste places, old rice paddies, along streams and ditches, roadsides, about dwellings, in gardens, and in damp meadows. It has small heads of white flowers in the leaf axils; the leaf-form is variable.
PREPARATION	The younger parts of the plant may be cooked and eaten as greens.

Ceylon spinach, Alugbati

Basella alba L. BASELLACEAE

DESCRIPTION This fleshy, twining vine grows in hedges and along fences. Sometimes it is cultivated. The rather fleshy stems may be dark red, purple, or yellowish green and the leaves may be green, red, or purplish. It does not occur in the forest but chiefly near settlements. The small flowers are pink and the fruits black or dark purple.

PREPARATION The whole plant may be eaten raw or cooked.

Hairless clearweed

Pilea glaberrima (Blume) Blume   URTICACEAE

DESCRIPTION   This erect plant, a somewhat juicy herb, grows 2 to 5 feet
               high. It has opposite leaves and numerous small greenish
               or greenish white flowers. A number of closely allied
               species occur, often in great abundance, in wet or damp
               high forests, in shaded ravines, and along streams, but
               always in the forest.

PREPARATION   In Java the tender young leaves and stems are eaten both
               raw and cooked, and are actually sold in the native
               markets.

Purslane

Portulaca oleracea L.   PORTULACACEAE

DESCRIPTION	This very fleshy weed-like plant is often abundant in settled areas throughout the Tropics, while other forms occur near the sea.
PREPARATION	The whole plant may be eaten raw or cooked as greens.

Seaside purslane

Sesuvium portulacastrum (L.) L.   AIZOACEAE

DESCRIPTION   This is a trailing branched herb with fleshy stems and
leaves, occurring only back of the beach or brackish
marshes along the shores of lagoons, etc., within the
influence of salt or brackish water. Widely distributed in
all tropical countries.

PREPARATION   The whole plant may be eaten raw or cooked as greens, but
it is desirable to change the water two or three times to
eliminate the salt.

Punarnava, Red spiderling

Boerhaavia diffusa Brandegee NYCTAGINACEAE

DESCRIPTION	This is a rather diffuse, spreading or ascending, branched herb with small pink flowers, the stems often reddish or purplish. Common in open places especially near the seashore back of the beach.
PREPARATION	The somewhat thickened leaves and young somewhat fleshy stems may be cooked and eaten.
WARNING	The roots are reported as being eaten in Fiji in times of scarcity, but as their use as food affects the kidneys, they should be used with caution if at all.

Black nightshade

Solanum americanum Mill.   SOLANACEAE

SYNONYM	*Solanum nigrum* L.
DESCRIPTION	Thus is an erect, branched herb that normally grows 1 to 2 feet high. It has small white flowers and small black berries. It is common both in waste places and cultivated lands.
PREPARATION	The young leafy shoots make excellent greens when cooked, and are extensively so used in the Tropics of both hemispheres. The small black fruits are edible.

Swamp morning-glory

Ipomoea aquatica Forssk. **CONVOLVULACEAE**

DESCRIPTION	This vine grows only in shallow fresh water ponds and swamps, and resembles the sweet-potato vine. It has pink flowers.
PREPARATION	The tender stems and young leaves make very excellent greens, and are frequently gathered and sold in the native markets for this purpose.

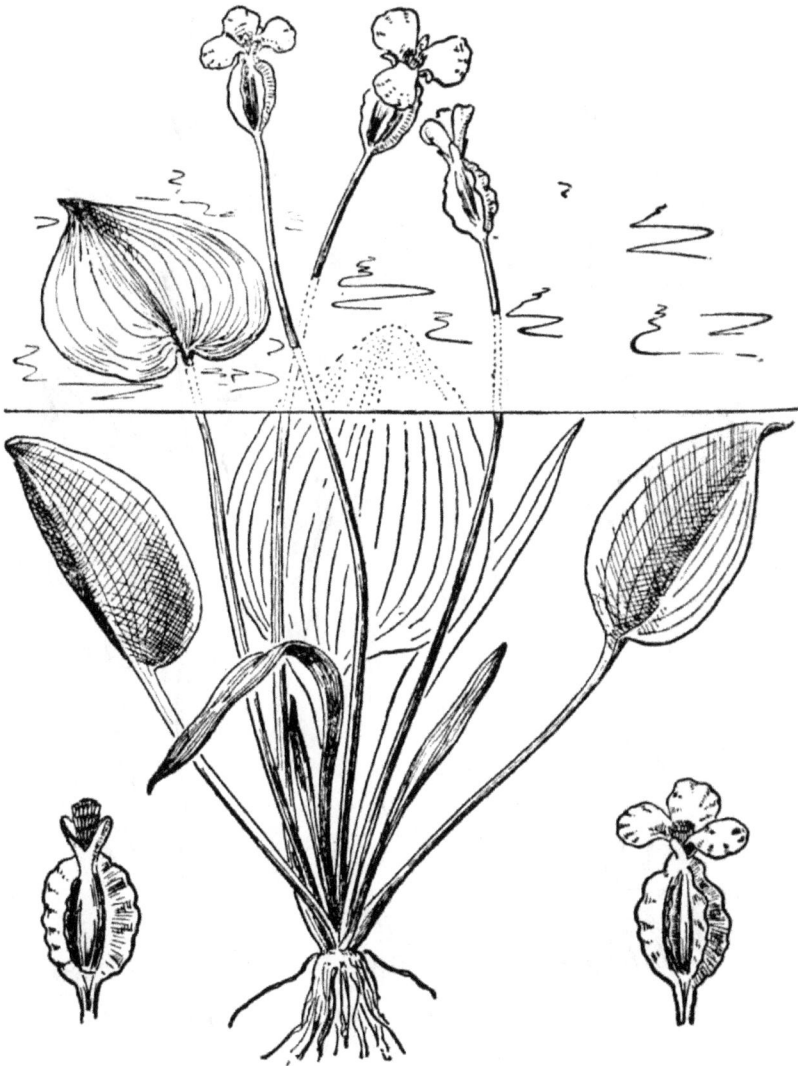

Duck-lettuce

Ottelia alismoides (L.) Pers.   HYDROCHARITACEAE

DESCRIPTION   This herb has rather large, thin leaves. It grows in slow shallow streams, pools, and quiet ponds, the white flowers extending above the surface of the water, the leaves wholly submerged or extending to the surface.

PREPARATION   The entire plant may be cooked and eaten as greens.

(A) *Monochoria vaginalis*, (B) *Monochoria hastata*

False pickerelweed

Monochoria vaginalis (Burm.f.) C.Presl PONTEDERIACEAE
Monochoria hastata (L.) Solms

DESCRIPTION	This herb is somewhat fleshy and grows about 1 to 1-1/2 feet high. It grows in open wet places, old rice paddies, or along streams, and is often abundant. It has blue flowers.
PREPARATION	The whole plant except the roots may be eaten raw, steamed, or boiled.

Water hyacinth

Eichhornia crassipes (Mart.) Solms   PONTEDERIACEAE

DESCRIPTION   While this plant is a native of Brazil and of comparatively recent introduction in the Old World Tropics, it is now widely naturalized and wherever it occurs it is usually very abundant. The flowers are blue with a yellow spot. They float on ponds and slow streams, and also occur as a weed in rice paddies.

PREPARATION   In Malaya the young leaves, leaf stalks, and flowering parts are steamed or boiled and eaten.

Lilac tasselflower

Emilia sonchifolia (L.) DC. ex DC.   ASTERACEAE

DESCRIPTION   This is a common and widely distributed weed and usually grows less than 1 foot high. It is found in open places, meadows, wastelands, and coconut plantations, but not in the forest. The flowers are pink or some forms, reddish. It is botanically allied to lettuce.

PREPARATION   The whole plant may be eaten raw cooked.

Fireweed

Erechtites spp.

ASTERACEAE

DESCRIPTION   These two weeds are very abundant in deserted clearings and waste places. *Erechtites valerianifolia* (Link ex Wolf) Less. ex DC. (A) bears pink flowers; *Erechtites hieracifolia* (L.) Raf. (B), yellowish flowers. The plants are usually 2 to 3 feet high.

PREPARATION   The tender parts may be cooked and eaten as greens.

Toothache plant, Paracress

Spilanthes acmella (L.) L.   ASTERACEAE

DESCRIPTION   This weedy herb grows both erect or ascending. It is branched and bears yellow flowers. The plant grows abundantly in meadows, waste places, along paths, in abandoned agricultural lands, but not in the forests.

PREPARATION   The younger parts of the plant may be cooked and eaten as greens. Also used medicinally to treat toothaches.

Indian camphorweed

Pluchea indica (L.) Less. ASTERACEAE

SYNONYM	*Baccharis indica* L.
DESCRIPTION	This small shrub grows 2 to 3 feet high. It is common and widely distributed, especially near the seashore and in wet soil. It bears a pale violet flower.
PREPARATION	The young leaves, tips of the branches, and young flowers may be cooked and eaten, and are extensively so used in Java.

Indian acalypha

Acalypha indica L.   **EUPHORBIACEAE**

DESCRIPTION   This is an erect, branched herb which grows up to 3 feet high. It occurs as a weed about settlements, in meadows, along ditches, and in waste places generally, often abundantly; it does not occur in the forests.

PREPARATION   The young leaves and tender stems may be cooked and eaten.

Copperleaf

Acalypha wilkesiana Müll. Arg. EUPHORBIACEAE

DESCRIPTION This is an ornamental shrub which grows 5 to 15 feet high. It has green or reddish twigs and variegated leaves, green and variously mottled or light red, dark brownish red, sometimes with greenish yellow blotches, or pale edges (B). It is easily recognized by its colored leaves. It is a native of Polynesia and planted in hedges and near houses throughout Malaya, frequently abundant.

PREPARATION The young shoots and young leaves may be cooked and eaten. There are various other species of this genus, herbs, shrubs, or small trees, all with green leaves, whose young parts may also be similarly prepared and eaten with safety.

Malunggai, Moringa, Horseradish tree

Moringa oleifera Lam.   MORINGACEAE

DESCRIPTION   This is a small or medium-sized tree, 15 to 20 feet high, with fine thin compound leaves and white flowers. It is cultivated and spontaneous in many parts of the Old World Tropics, but is not found in the forests.

PREPARATION   The leaves, shoots and young pods (sometimes termed "drumsticks") make excellent greens when cooked, or they may be eaten raw. The roots have the characteristic biting taste of horseradish. The mature pods are too tough to be eaten, but the seeds may be roasted and used as food.

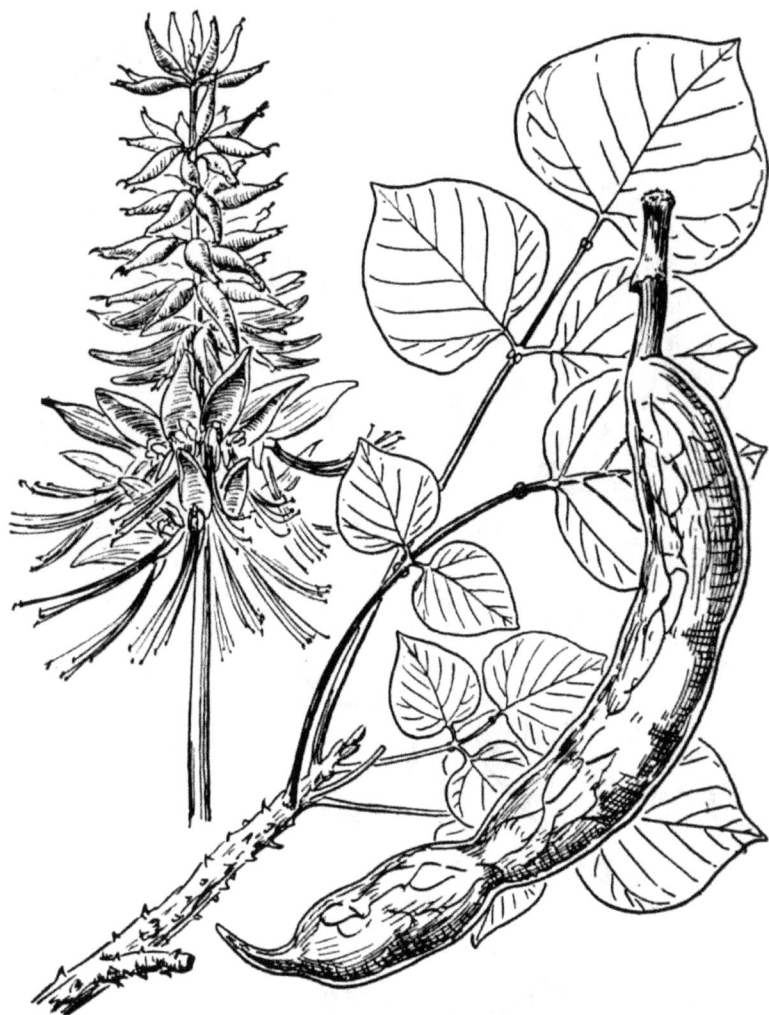

Coral tree

Erythrina variegata L.   FABACEAE

DESCRIPTION   This tree grows from 20 to 50 feet in height. It is common along the seashore and is often planted in settled areas along roadsides. The rather large, crowded flowers are bright red. It does not occur in the high forest. The young branches are usually somewhat spiny.

PREPARATION   The leaves and the tender shoots may be steamed or boiled and eaten as greens.

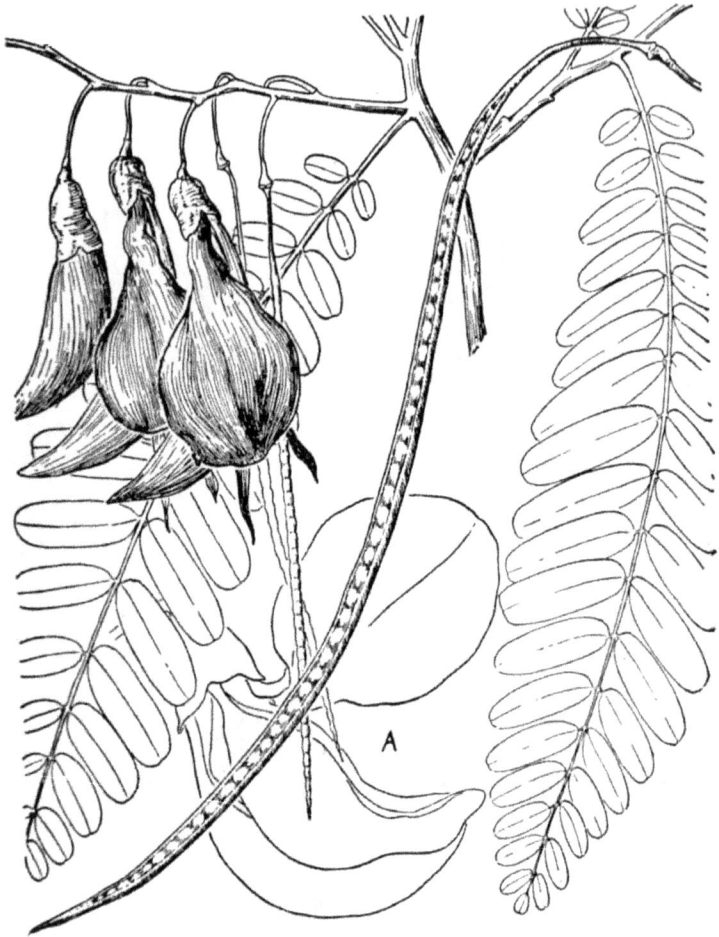

Agati, Hummingbird tree

Sesbania grandiflora (L.) Pers.   FABACEAE

DESCRIPTION	This slender tree bears long, slender, hanging pods up to 2 feet long or longer, and large white or wine-red flowers, these 2 to 3 inches long. These trees are sometimes planted and often naturalized; they are not in the forests. An outline of the very large flower is shown in (A).
PREPARATION	The young leaves and the young pods may be cooked and eaten, while the large flowers and flower buds are very commonly cooked and used as food.
WARNING	Do not eat the mature seeds.

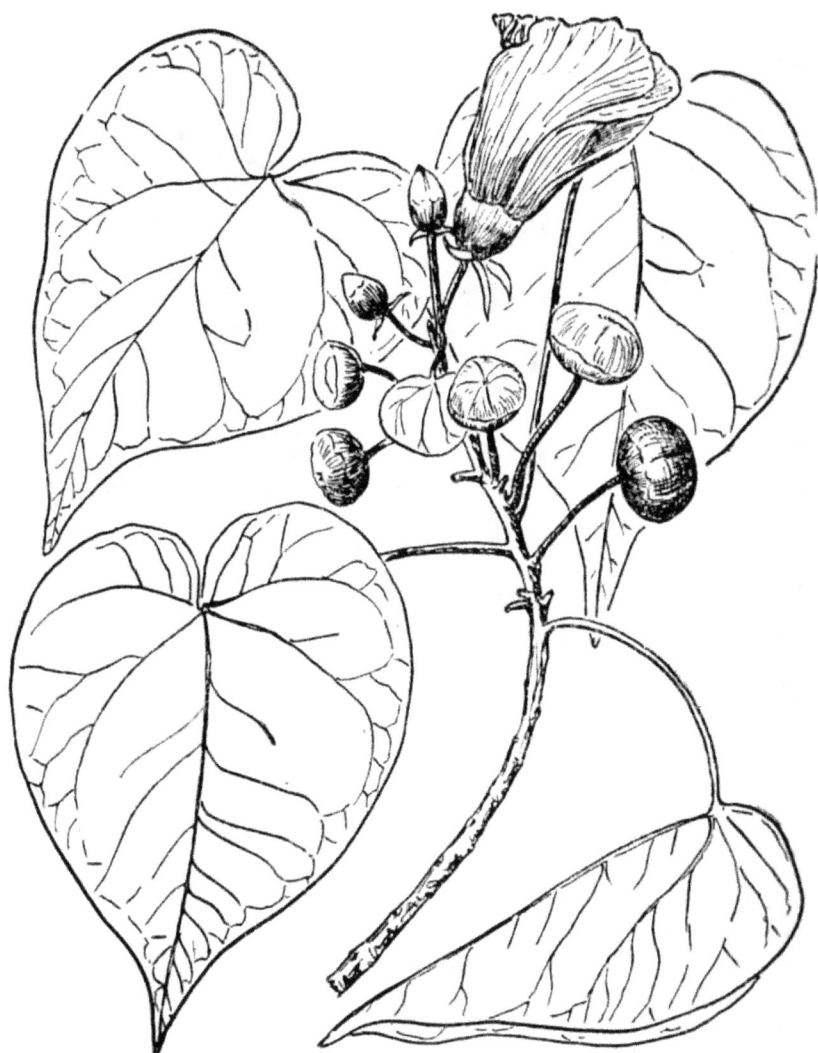

Portia tree

Thespesia populnea (L.) Sol. ex Corrêa   MALVACEAE

DESCRIPTION   This is a small or medium-sized tree bearing large yellow flowers. It is found chiefly near the seashore and usually immediately back of the beach.

PREPARATION   The smooth leaves may be eaten raw or cooked, as well as the flower buds and the flowers. The rather dry, nearly, round fruits are not edible.

Pemphis

Pemphis acidula J.R. Forst. & G. Forst.   LYTHRACEAE

DESCRIPTION	This is a small tree attaining a height of 10 or 12 feet. It has small, 6-parted flowers and small leaves. It is found only along the seashore, where it is common and very widely distributed.
PREPARATION	The small leaves have a distinctly acid taste and may be eaten raw.
NOTE	A popular bonsai plant.

Tree heliotrope

Tournefortia argentea L. f.　　　　　　　　BORAGINACEAE

SYNONYM	*Heliotropium foertherianum* Diane & Hilger
DESCRIPTION	This is a shrub with stout branches. It has grayish white, very hairy leaves and many small, crowded flowers. It grows only on sandy seashores, and is common and widely distributed.
PREPARATION	The leaves may be eaten raw.

Indian mulberry, Cheese fruit, Noni

Morinda citrifolia L. RUBIACEAE

DESCRIPTION This is a shrub or small tree, varying from 4 or 5 to 10 or 15 feet in height. It is common along the seashore. The flowers are white and the fruits are greenish white.

PREPARATION The young leaves and the young fruits may be eaten raw or cooked, and the mature fruits, deprived of their seeds, may also be eaten.

Cantala

Agave cantala (Haw.) Roxb. ex Salm-Dyck   ASPARAGACEAE

DESCRIPTION	This form of the "century plant" is extensively grown in the drier parts of the Malayan region, sometimes in plantations, sometimes as scattered plants in fencerows, waste places, etc., where it is naturalized. The thick, fleshy leaves, 3 to 6 feet long, are very sharp-pointed and their edges are very spiny.
PREPARATION	In Java, the tender heart or "cabbage" (A) in the crown of the growing plant is extensively eaten. It should be cut into small pieces and well cooked, preferably with one or two changes of water.
WARNING	Many of the American species are not edible; some contain saponin and others contain minute stinging crystals of oxalate of lime. Seek the advice of natives whenever possible.

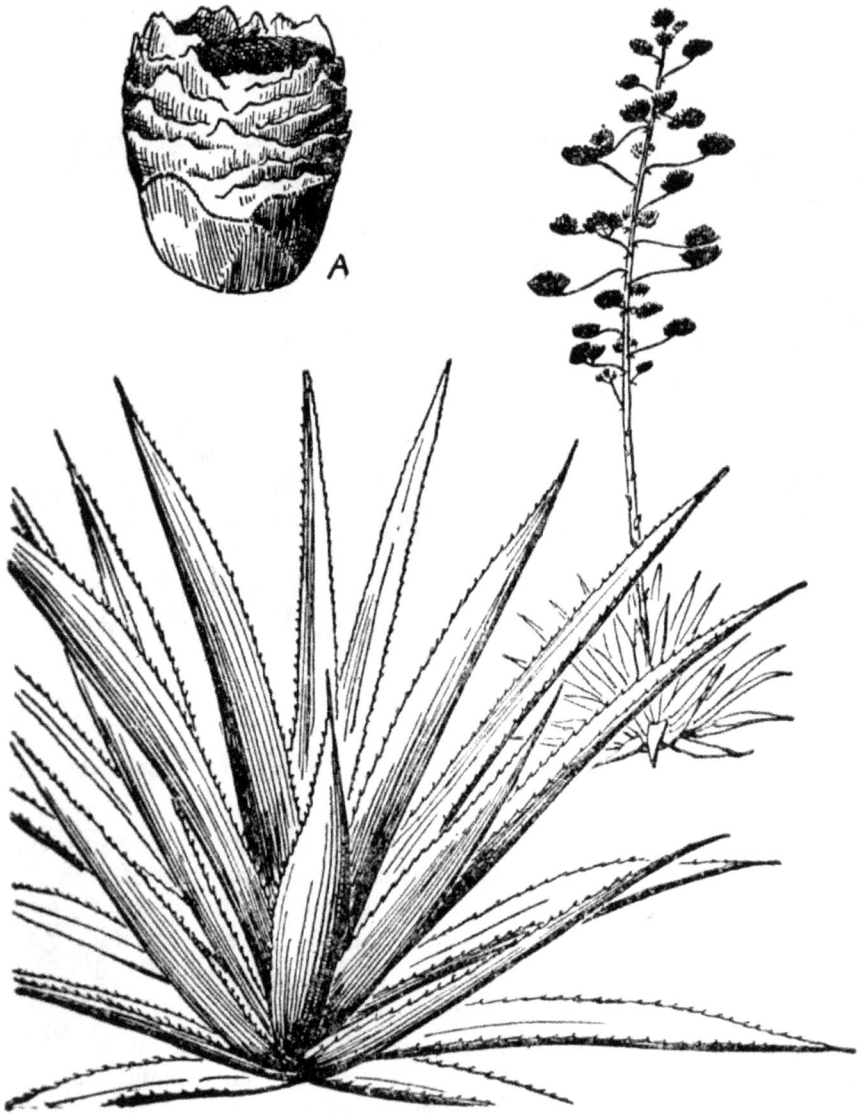

Cantala *(Agave cantala)*
(A) heart or "cabbage"

Edible Fruits

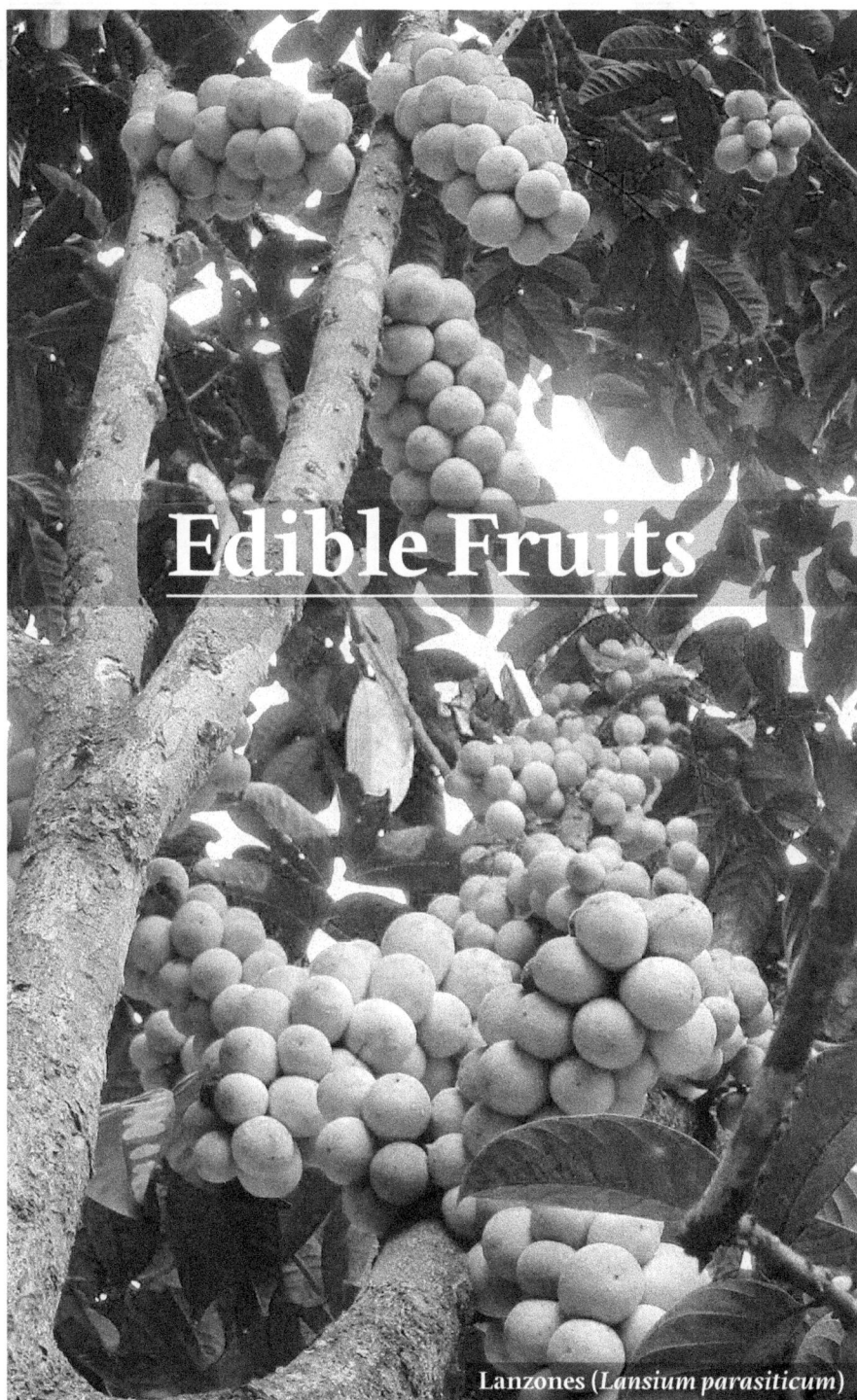

Lanzones (*Lansium parasiticum*)

Edible Fruits

Bananas in General

There are scores of varieties of the common banana, and for all practical purposes the plantain cannot be distinguished from the edible banana except that all the bananas may be eaten raw, while the plantain fruits must be cooked–either boiled, fried, or roasted; green and ripe bananas may also be cooked. The fruits vary much in shape and size as well as in color, varying at maturity from green to various shades of yellow, or even brownish purple. Many wild forms occur in the forested regions (usually, however, not in the high forest except along streams). The fruits of the wild species contain numerous seeds and small quantities of pulp, but even these may be gathered when young and cooked.

Other parts of the banana plant may be, used as food, especially the fairly large, more or less cone-shaped, terminal flower bud (see A, next page). These flower buds may be boiled or roasted in hot ashes, and certain varieties make an excellent vegetable; others contain fairly large quantities of tannic acid and are hence bitter, but the bitter principle may be eliminated in part by cooking in several changes of water. With the bitter kinds it is best to cut the bud into rather small pieces before cooking.

The soft inner parts of the rather thick root and the tender heart of the base of the stems may be cut into small pieces, boiled, and eaten. Even the small shoots from the lower parts of the plant may be cooked and eaten when nothing better is available.

In general these statements apply to all types of the banana, whether wild or cultivated. While the parts other than the fruits and the flower buds do not rate as first class food by any means, they are safe to eat when boiled or roasted.

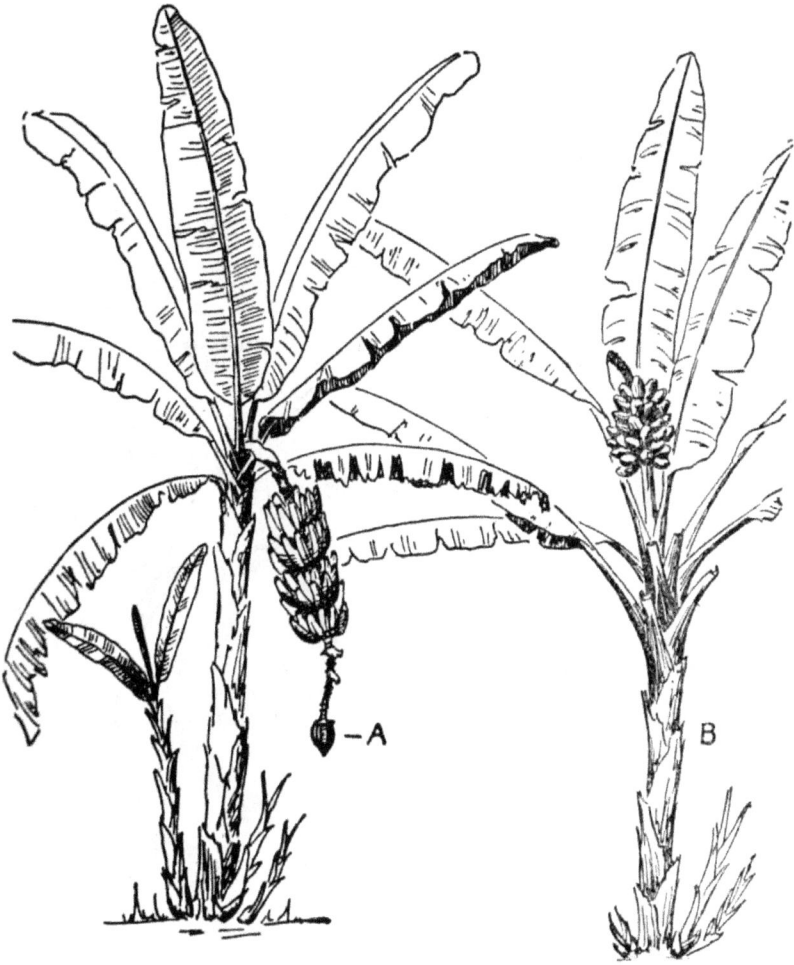

(A) *Musa × paradisiaca* L.; (B) *Musa troglodytarum* L.

Banana, Saging

Musa spp.

MUSACEAE

DESCRIPTION This fruit is too well known to discuss here. The banana may be eaten raw or cooked, but the plantain requires cooking. Other parts of the cultivated bananas and plantains, and of the wild forms that occur in the forests and old clearings, may be boiled or roasted and eaten, especially the large flower bud (A). (B) is a wild and cultivated banana extending from the Moluccas and New Guinea to Polynesia, with erect fruit clusters.

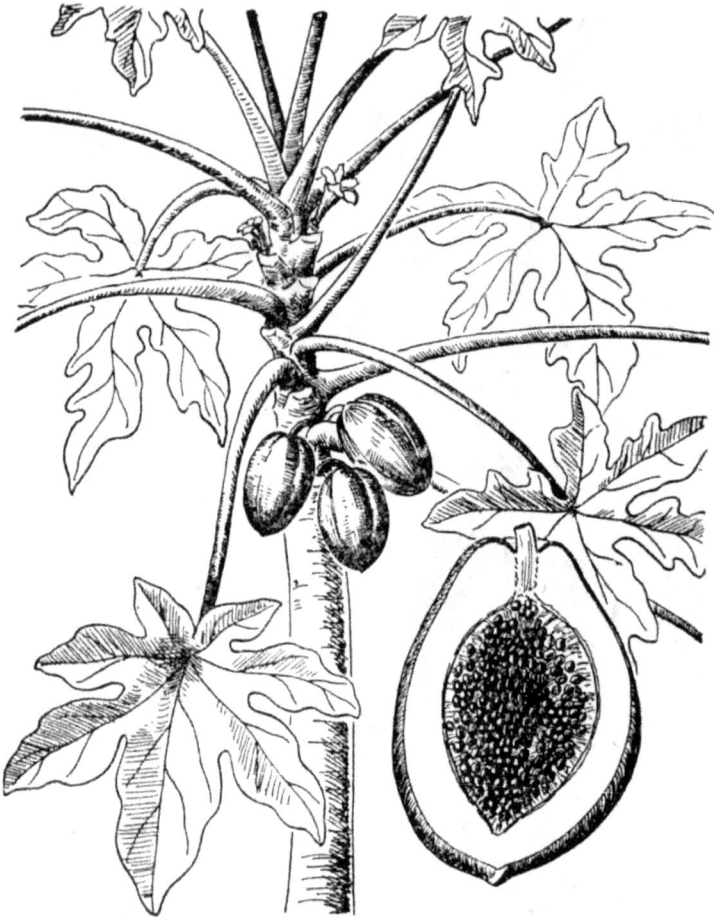

Papaw, Papaya, Kapayas

Carica papaya L.   CARICACEAE

DESCRIPTION	This is a soft wooded, erect, normally unbranched tree. It usually grows from 6 to 15 feet high. The large yellow melon-like fruits are borne on the trunk, and are excellent food.
PREPARATION	The green immature fruits may be cooked and eaten. The young leaves and leaf-stems and flowers (the male flowers borne on separate plants) may be cooked and eaten as greens. It is, however, important that such parts be cooked with several changes of water to remove the bitter taste and certain harmful substances.

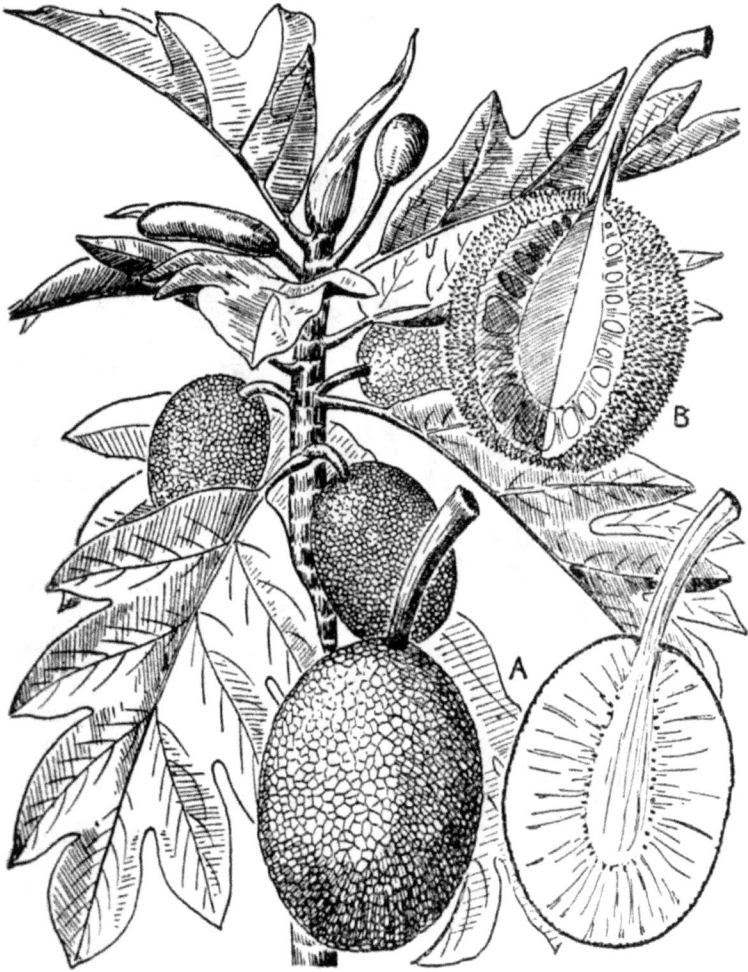

Breadfruit, Kamansi

Artocarpus altilis (Parkinson ex F.A.Zorn) Fosberg MORACEAE

DESCRIPTION	This is a large tree that may grow 40 feet high or more. It has very large lobed leaves and rather large, nearly round green or brownish green fruits.
PREPARATION	It is a basic food plant in many parts of Polynesia, where many cultivated varieties are found. The seedless form (A) is utilized either boiled, baked, or fried. The large seeds in the seeded form (B, Malaysia), boiled or roasted, are excellent food, as are the seeds of other species.

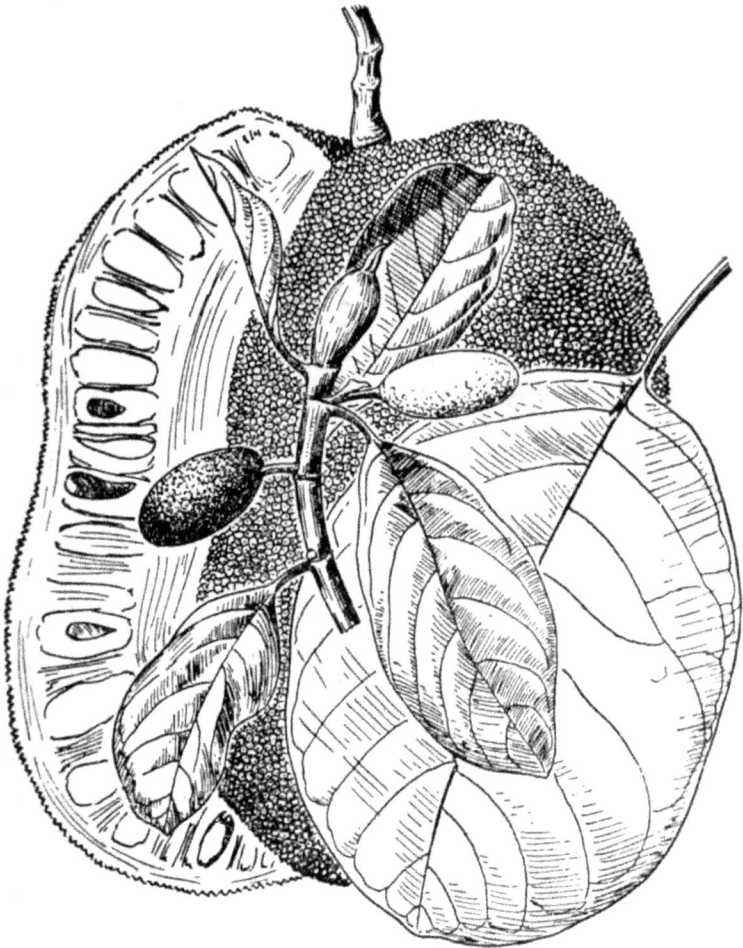

Jack fruit, Langka, Nanka

Artocarpus heterophyllus Lam.   MORACEAE

DESCRIPTION   This is a large tree, normally cultivated only. The very large greenish or yellowish green fruits are 1 to 3 feet long, and sometimes weigh up to 40 pounds. The fruits are borne directly on the tree trunk and larger branches. There are many different species of *Artocarpus* in Malaysia. All have abundant milky sap, and the seeds of all of them are edible when cooked.

PREPARATION   The pulp may be eaten raw. The numerous large seeds make excellent food when boiled or roasted.

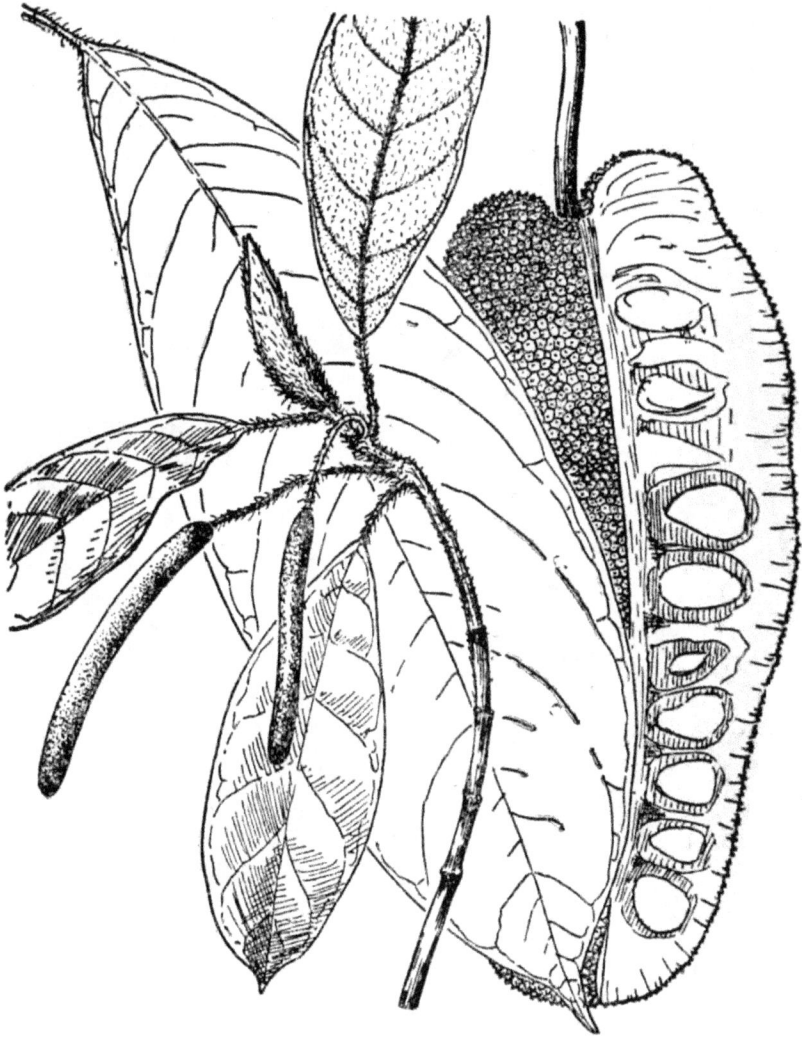

Champedak

Artocarpus integer (Thunb.) Merr.   MORACEAE

DESCRIPTION   This large tree has a milky sap. Its leaves are more or less hairy. The large, cylindric fruits borne on the larger branches are smaller than in the jack fruit and have a very strong odor.

PREPARATION   The pulp is edible and the seeds are edible when boiled or roasted.

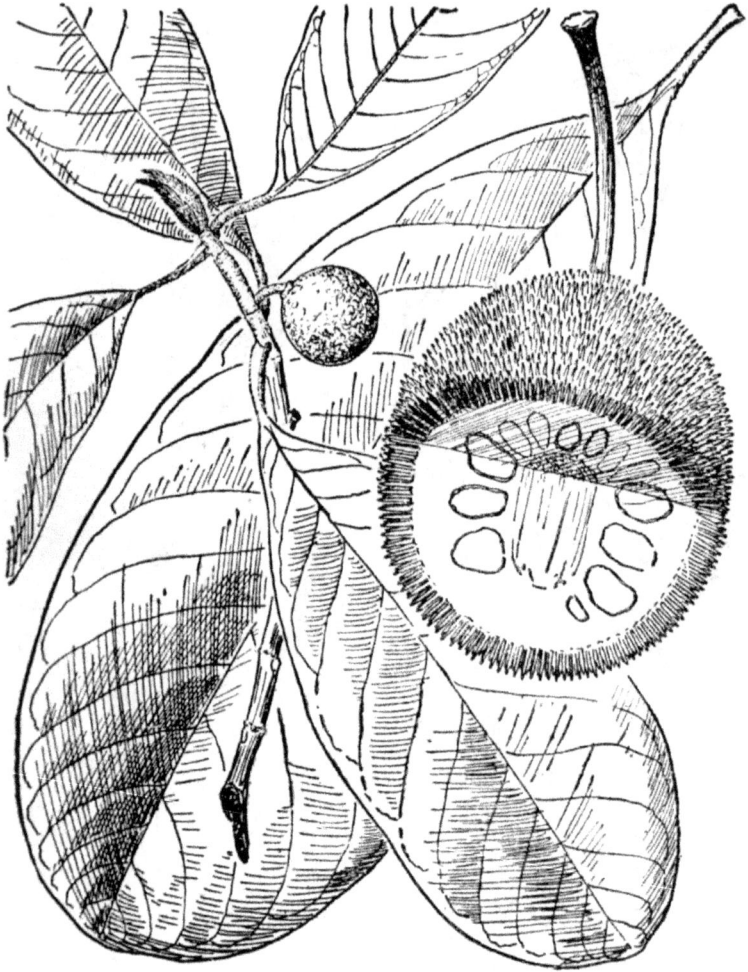

Monkey jackfruit

Artocarpus rigidus Blume   MORACEAE

DESCRIPTION   This large tree has a milky juice. The fruit is round,
               greenish or greenish brown, and is found on the smaller
               branches. These fruits are up to 5 inches in diameter and
               covered with short stiff conical spines. This tree occurs in
               Malaysia, but not in Polynesia.

PREPARATION   The fruit is one of the best of Malayan fruits, the well-
               flavored pulp being eaten raw. The large seeds should be
               boiled or roasted before eating.

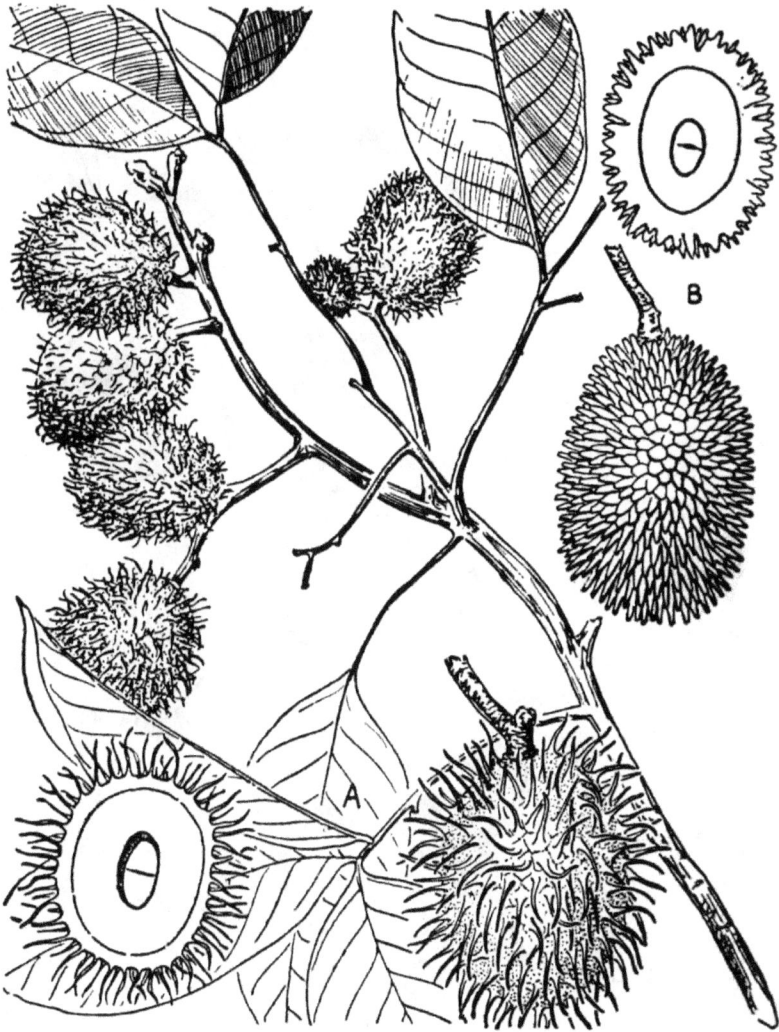

Rambutan (A), Pulusan (B)

Nephelium lappaceum L. (A)   SAPINDACEAE
Nephelium mutabile Blume (B)

DESCRIPTION   These fairly large trees are usually cultivated, but the
                     Pulusan is sometimes found in forests. The characteristic
                     red fruits are one-seeded. The seed is surrounded by a
                     white unusually well-flavored pulp. Both the Rambutan
                     and the Pulusan are among the very best fruits of Malaya.

Lanzone

Lansium parasiticum (Osbeck) K.C.Sahni & Bennet   MELIACEAE

SYNONYM	*Lansium domesticum* Corrêa
DESCRIPTION	This a cultivated tree. The pale, yellowish fruits are found on the trunk and on the larger branches below the leaves. It is one of the best of the tropical fruits.

Guava, Bayabas

Psidium guajava L.   MYRTACEAE

DESCRIPTION   This is a small shrub or tree that grows 5 to 15 feet high, often abundant, but never found in real forests. It has white flowers and pale greenish or yellowish green, smooth, many-seeded fruits.

PREPARATION   This excellent fruit may be eaten raw or cooked.

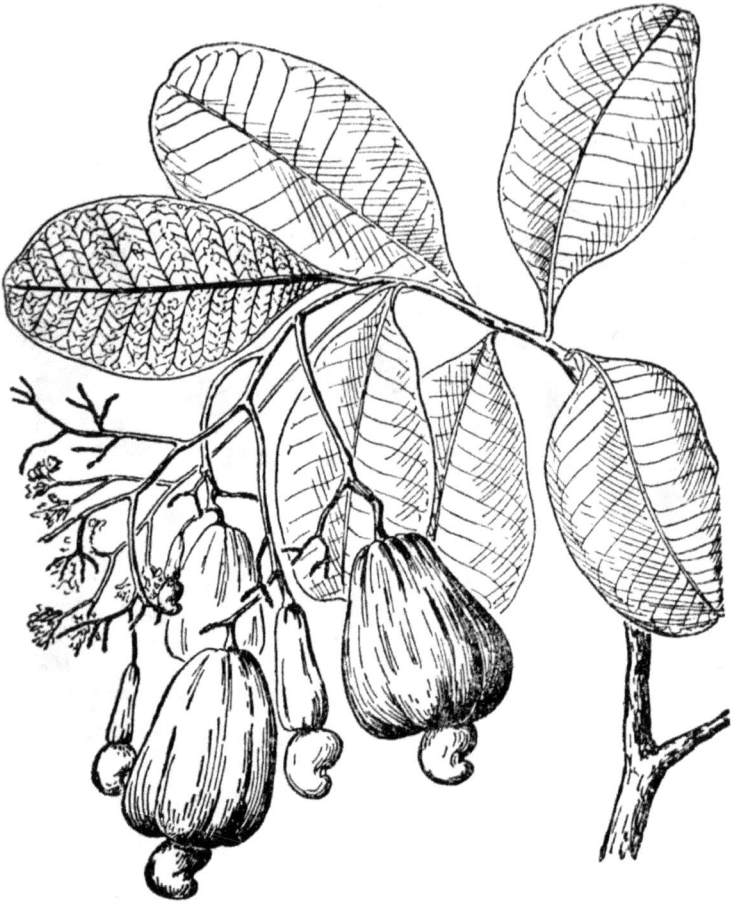

Cashew

Anacardium occidentale L. ANACARDIACEAE

DESCRIPTION	This is a small or medium-sized tree usually about 20 feet high. ft is often common in more or less settled areas, but not in the forests.
PREPARATION	The yellowish to purplish very juicy large part of the fruit is very refreshing. The single seed in the smaller part of the fruit is the cashew nut of commerce and should be eaten boiled or roasted.
WARNING	The sap in the shell of the small part of the fruit surrounding the seed is very caustic. In boiling or roasting the seed-bearing part, avoid the steam or smoke.

Sweet sop

Annona squamosa L.   ANNONACEAE

DESCRIPTION   This is a small tree, usually 15 feet high, and is found both wild and in cultivation. This tree is found chiefly in and near settlements, not in the forests.

PREPARATION   The medium-sized pale green fruit is of excellent flavor and is always eaten raw.

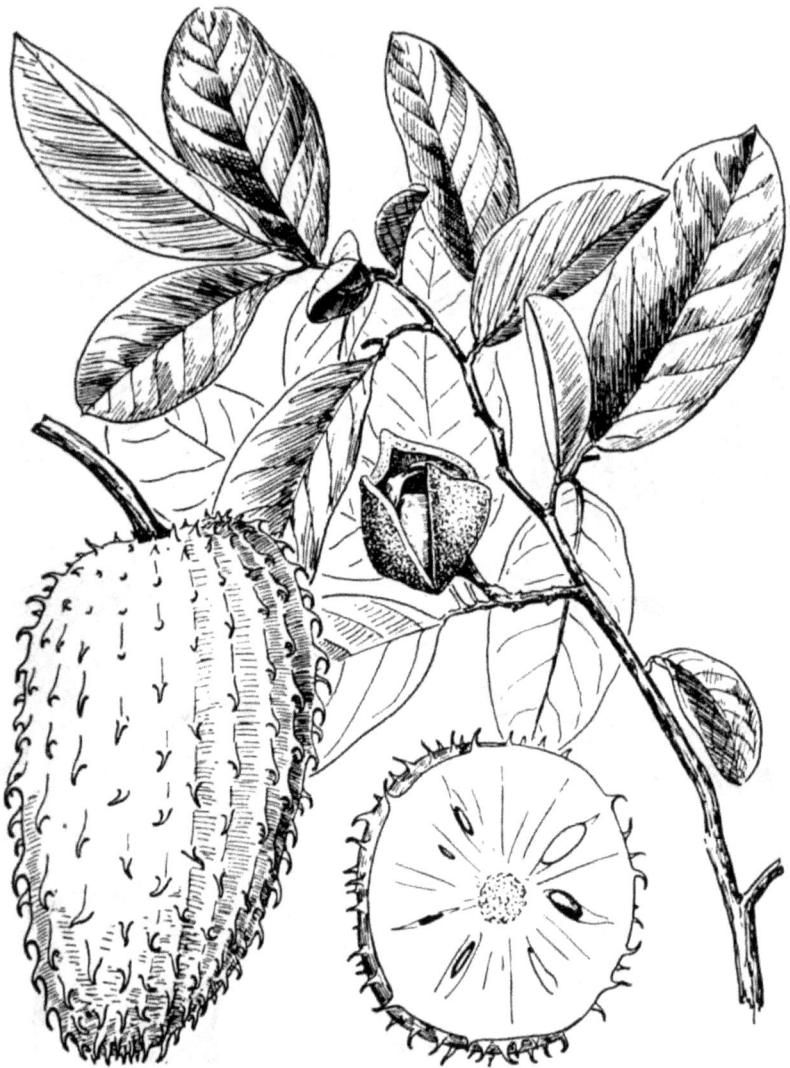

Sour sop, Guyabano

Annona muricata L.   ANNONACEAE

DESCRIPTION   This tree is about 15 feet high, rather similar to the sweet
sop, is generally cultivated, but is sometimes wild. It is
found chiefly in and near settlements, not in the forests.

PREPARATION   The large, well-flavored greenish fruits are always eaten
raw.

Custard apple

Annona reticulata L.   ANNONACEAE

DESCRIPTION   This tree, about 15 feet high, is similar to the sweet sop and the sour sop. It is found chiefly in cultivation in the settled areas, never in the forests.

PREPARATION   The large well-flavored greenish juicy fruits are eaten raw.

Mango

Mangifera indica L.   ANACARDIACEAE

DESCRIPTION	This is usually a large tree, mostly planted; not found in the forests. Most of the varieties in Malaya and Polynesia have yellow fruits.
PREPARATION	It is one of the very best of tropical fruits.
WARNING	Rarely an individual may be allergic to mangoes, and in such cases a skin rash may develop; very rarely the individuals may be affected by the leaves. Other species of the genus in Malaya, all with edible fruits, have a very irritating sap affecting the skin quite as does poison ivy. The indicated treatment is the same as for poison ivy poisoning.

Sapodilla, Chico

Manilkara zapota (L.) P. Royen SAPOTACEAE

SYNONYM	*Achras zapota* L.
DESCRIPTION	This is a medium-sized tree, usually 15 to 25 feet high, with a milky sap. It grows both cultivated and spontaneously, but not in the forests.
PREPARATION	The grayish to brownish fruits vary in shape from round to oval and are excellent to eat when ripe. The pulp which is pinkish white to reddish brown is sweet and somewhat granular. The pulp surrounds several fairly large smooth black seeds. They should not be cooked.

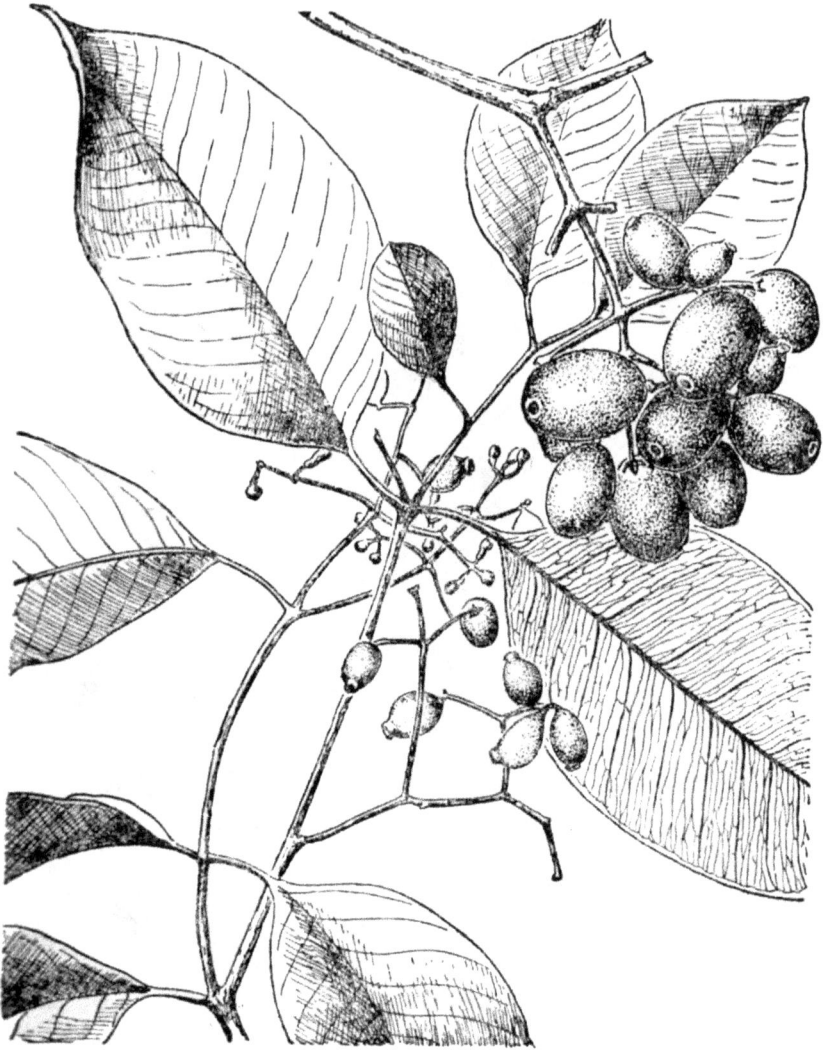

Jambolan, Duhat

Syzygium cumini (L.) Skeels   MYRTACEAE

DESCRIPTION This is a medium-sized tree which grows 20 to 30 feet high. It is found both wild and cultivated. The tree has somewhat leathery leaves, small white flowers, and one-seeded, light to dark purple, smooth fruits.

PREPARATION The single seed is surrounded by a whitish or yellowish, sourish, rather pleasant tasting edible pulp.

Water cherry, Lau lau

Syzygium aqueum (Burm.f.) Alston   **MYRTACEAE**

DESCRIPTION   This small tree, 15 to 20 feet high, is chiefly cultivated, but sometimes grows wild. The leaves are short-stalked and grow in opposite pairs. The flowers of all species are white, pink, or red, and always have many stamens. The edible fruits are smooth, pink in color, and somewhat juicy. The fruits vary greatly in size, some of them being dry with no pulp, others with fairly ample pulp, which is usually acid.

PREPARATION   There are many different species of *Syzygium* in the forested regions and their fruits may be eaten with entire safety, although some have almost no pulp.

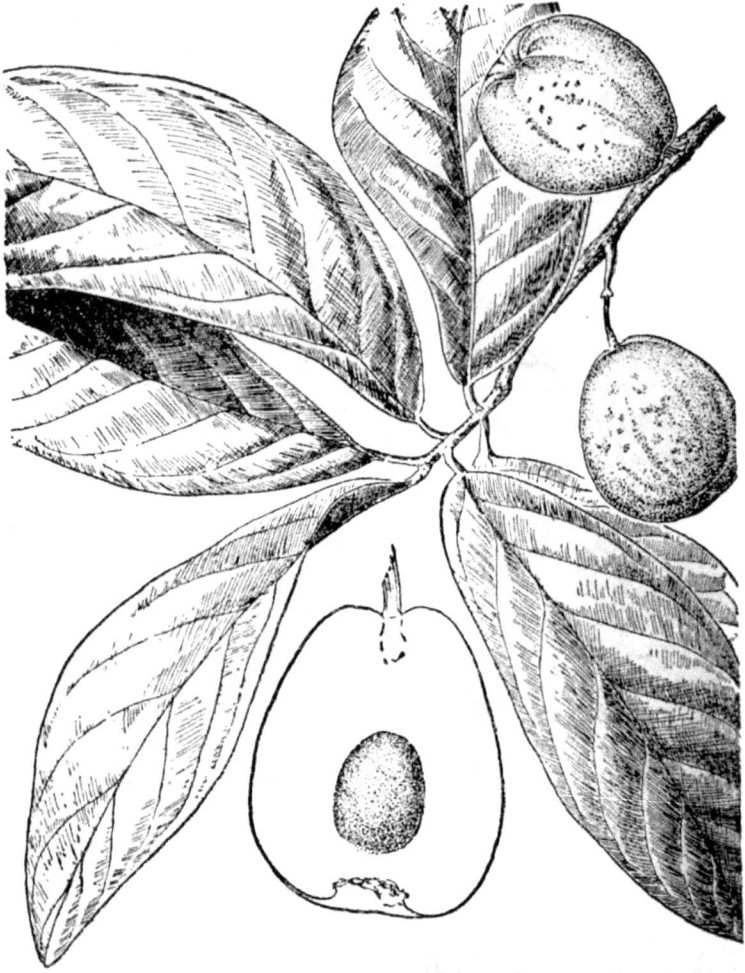

Malay apple

Syzygium malaccense (L.) Merr. & L. M. Perry MYRTACEAE

DESCRIPTION
: This is a medium-sized tree, 15 to 30 feet high, with somewhat leathery leaves. The red flowers are found on the branches below the leaves. This tree is chiefly cultivated. The pink to reddish, thin skinned, smooth fruit, somewhat resembling an apple, varies from 2 to 4 inches in length.

PREPARATION
: The thick, rather well-flavored pulp surrounding the large seed is edible.

Rose apple

Syzygium jambos (L.) Alston   MYRTACEAE

DESCRIPTION   This is a small tree, 10 to 15 feet high, and has white
flowers. The fruits are somewhat rose-scented, greenish
white, egg-shaped or somewhat pear-shaped and are about
1 inch long. This tree is often planted and occurs also in
thickets, waste places, and secondary forests. The tree has
a wide distribution.

PREPARATION   The fruits are eaten raw.

Santol

Sandoricum koetjape (Burm.f.) Merr   MELIACEAE

DESCRIPTION This is a medium-size tree, about 30 feet high, and bears round, yellowish fruits about 2 to 5 inches in diameter. It grows both wild and cultivated and is often rather common. The fruits are covered with very short hairs and contain from two to five fairly large seeds, surrounded by a dirty white, soft, juicy, sour-sweet, edible pulp.

PREPARATION The seeds are not eaten, only the surrounding pulp.

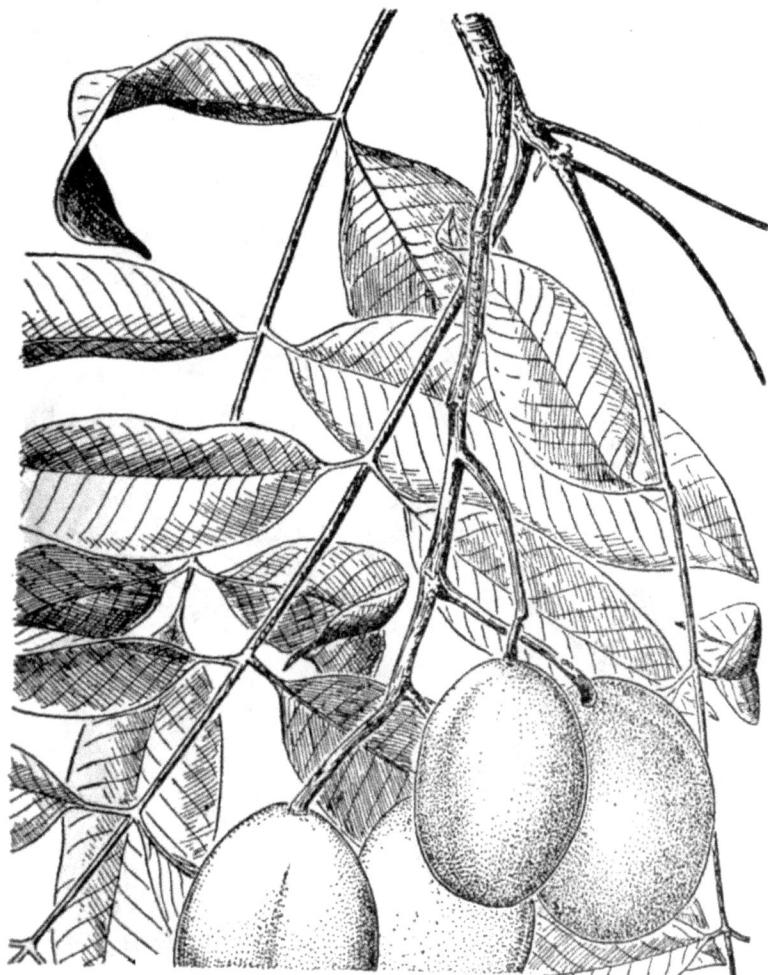

Polynesian plum

Spondias dulcis Parkinson ANACARDIACEAE

DESCRIPTION This tree, 25 to 30 feet high, is widely distributed in Polynesia. It is often planted. The fruits are plum-like and art yellowish or yellowish green. While this species occurs also in some parts of Malaysia as a cultivated tree, its place in the forests is taken by a very similar species, *Spondias pinnata* (L. f.) Kurz, the fruits of which are also edible.

PREPARATION The thin pulp surrounding the large seed-bearing part is excellent to eat.

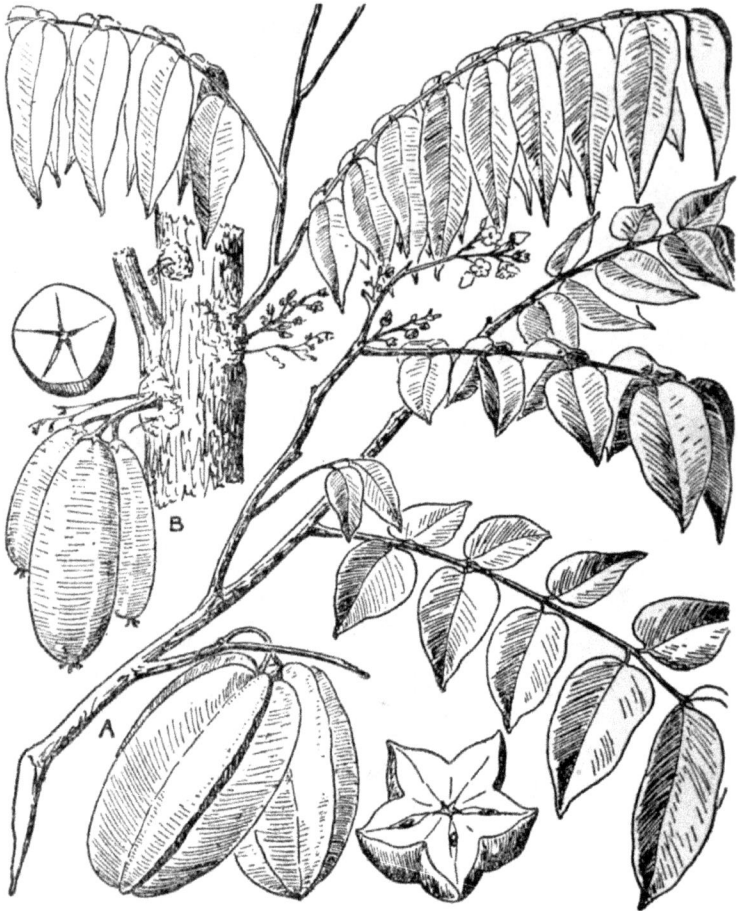

(A) Carambola (*Averrhoa carambola*); (B) Bilimbi (*Averrhoa bilimbi*)

Bilimbi and Carambola

Averrhoa bilimbi L. OXALIDACEAE
Averrhoa carambola L.

DESCRIPTION These are small trees, 12 to 15 feet high, with green or pale green, very acid fruits which may be eaten raw or cooked. In one, the smooth fruits, similar to small green cucumbers, are borne on the trunks and larger branches; in the other, the fruits borne on the small branchlets are sharply five-angled and star-shaped in cross section. Cultivated and wild in the settled areas, not in the forests.

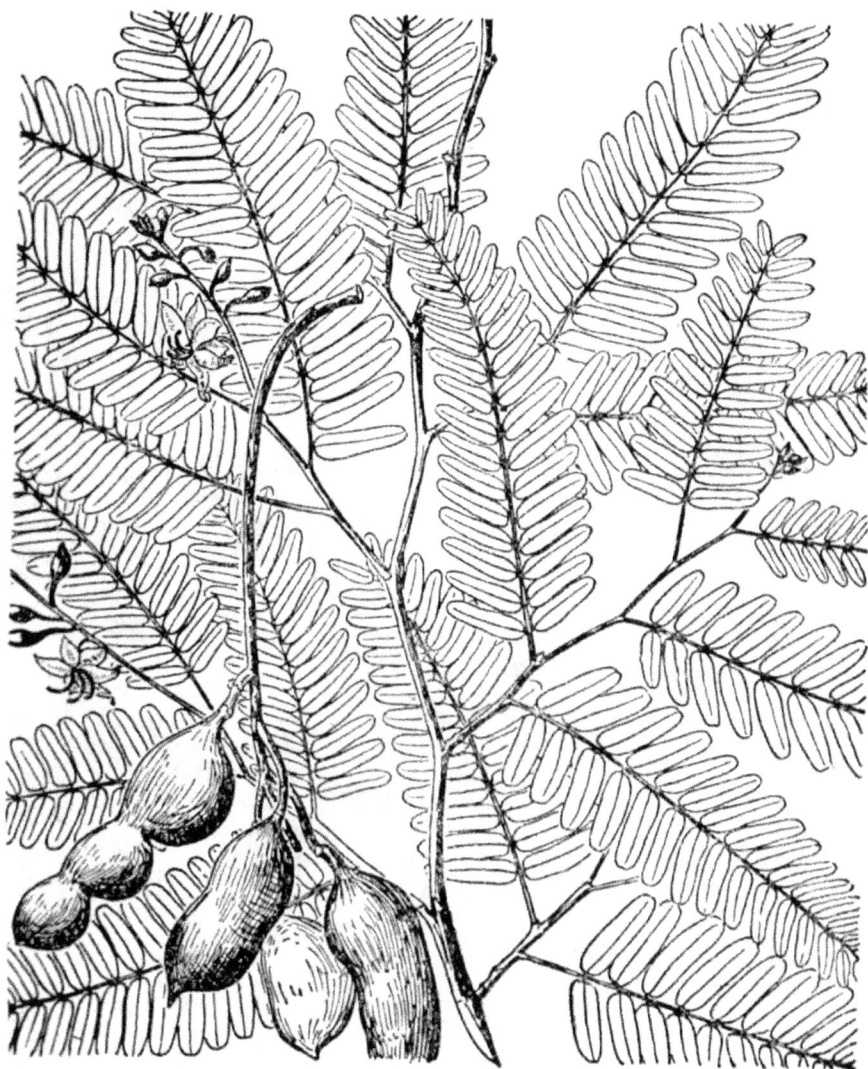

Tamarind, Sampalok

Tamarindus indica L.   FABACEAE

DESCRIPTION	This is a large tree, often planted, sometimes wild, but not found in the forests. The fruit is brown.
PREPARATION	The acid pulp surrounding the seeds may be eaten; it is a mild laxative. The young leaves and flowers may be cooked and eaten as greens.

Nam-nam

Cynometra cauliflora L.　　　　　　　　　　　　FABACEAE

DESCRIPTION　　　This is a small tree, growing from 8 to 12 feet high. The flowers and fruits are borne on tubercles on the trunk and larger branches. The fruits are usually ripe from August to November. When ripe, the one-seeded fruits are yellowish green or dirty yellow. The parts surrounding the seed are yellowish white, fragrant, juicy, edible and sweet-sour or sour in taste. The tree does not occur in the forests and is chiefly found only in cultivation.

Pandan, Screw pine

Pandanus tectorius Parkinson ex Du Roi   PANDANACEAE

DESCRIPTION   This is one of the most common plants in all of Polynesia and Malaysia, chiefly occurring near the sea and often forming dense thickets back of the beach. The trees are small, usually about 12 feet high. It may be identified by the prop roots on the trunk, or the long spiny leaves arranged spirally at the ends of the branches. These statements apply to all of the numerous species of this genus in the forests of Malaysia, Micronesia, and Polynesia.

PREPARATION   The terminal tender leaf-bud or "cabbage" may be eaten raw or cooked. The scanty red fruit pulp is also edible, as are the small seeds.

Gnemon tree, Daeking tree, Melinjo

Gnetum gnemon L.   GNETACEAE

DESCRIPTION	This is a small tree, 15 to 20 feet high, with glossy leaves and one-seeded red fruits. This is a forest tree, but it is sometime planted. There are several other species; all, however, are woody vines and are found in the Malayan forests. Their seeds may also be eaten.
PREPARATION	The seeds may be eaten raw, roasted, or boiled, and the young leaves make excellent greens.

Bignai

Antidesma bunius (L.) Spreng.   PHYLLANTHACEAE

DESCRIPTION   This is a small tree commonly found in open places and secondary forests. There are many other species of this genus in Malaya and a few in Polynesia, their fruits, all edible, are smaller than in bignai.

PREPARATION   The numerous small, usually purple-black, one-seeded fruits are edible.

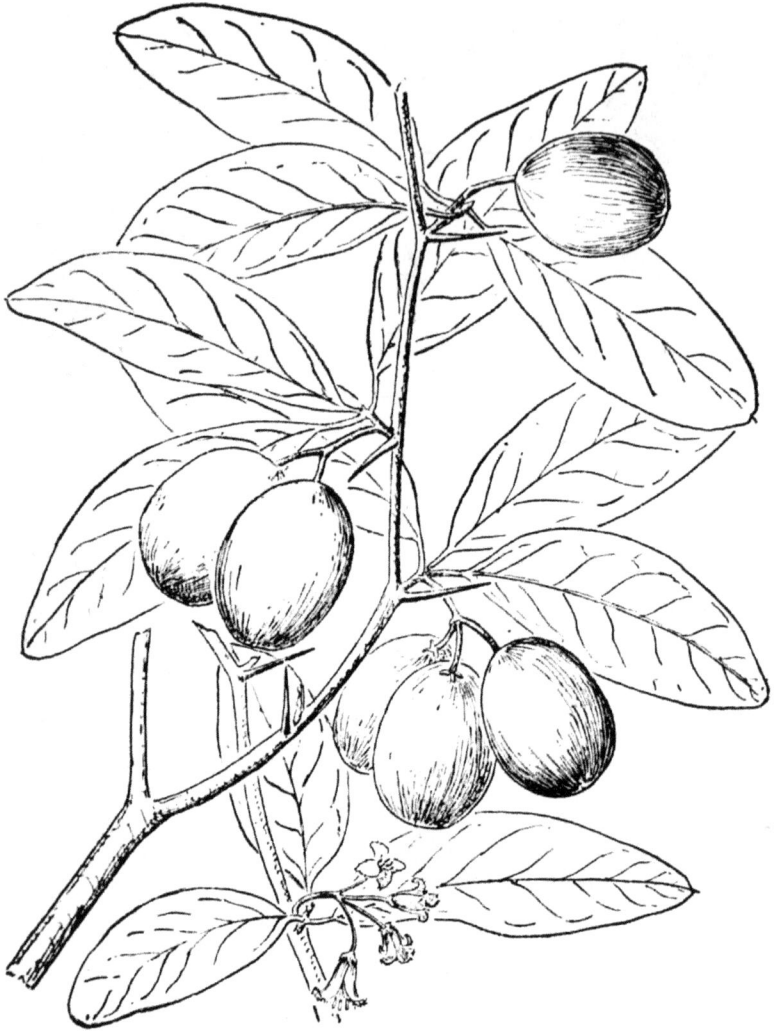

Tallow wood, Yellow plum, Sea lemon

Ximenia americana L.   OLACACEAE

DESCRIPTION   This is a small, spiny tree, always growing near the
                   seashore.

PREPARATION   The rather scanty sour pulp surrounding the large hard
                   seed-bearing part may be eaten raw, but the seeds should
                   not be eaten. The young leaves also may be cooked and
                   eaten.

Wild tomato

Lycopersicum esculentum Mill.   SOLANACEAE

SYNONYM	*Lycopersicon esculentum* Mill.
DESCRIPTION	This wild form of the common cultivated tomato is an erect, branched herb, 2 to 3 feet high, with leaves smaller than the cultivated form and small red fruit not larger than an English cherry. It is frequently common in deserted clearings, abandoned agricultural lands, and at higher altitudes even in open grasslands. It does not occur in the forests.
PREPARATION	The small red fruits are eaten raw.

Ground cherry

Physalis spp.   SOLANACEAE

DESCRIPTION	These may be identified as erect or ascending branched herbs with small white or yellow flowers. They are found in open waste places, sometimes near the seashore, but not in the forests.
PREPARATION	The round fruits are borne inside of an inflated husk. When mature, the fleshy round fruits are usually red. They somewhat resemble very small tomatoes and may be eaten raw.

Edible Seeds

Lotus seed pod (*Nelumbo nucifera*)

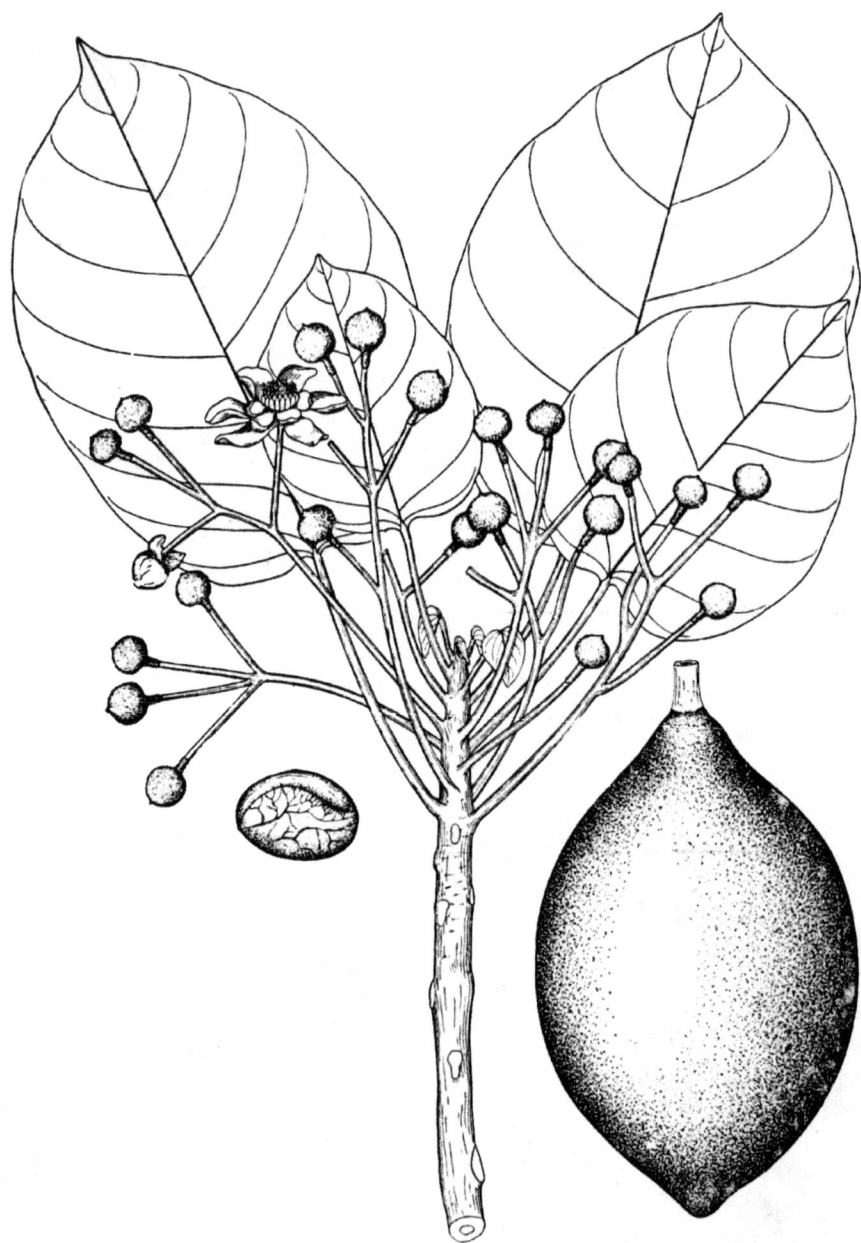

Pangi *(Pangium edule)*

Pangi

Pangium edule Reinw.   ACHARIACEAE

DESCRIPTION   This is a tree that grows 70 or 80 feet high. It is found in humid forests and also often planted. The large fruits (A) are up to 10 inches long, brown and densely rusty-hairy outside.

PREPARATION   The scanty pulp surrounding the numerous large hard seeds may be eaten when fully ripe.

WARNING   The seeds (B) are very poisonous (hydrocyanic acid). They are used as food by the natives but only after careful preparation. The seeds should be crushed and boiled for at least an hour, then put into running water for at least a day, after which they are boiled again and eaten; seek the assistance of competent natives if possible. The leaves also are poisonous if eaten.

Polynesian chestnut

Inocarpus fagifer (Parkinson) Fosberg   FABACEAE

DESCRIPTION   This is a small or medium-sized tree, 8 to 10 or 20 feet high, much more common in Polynesia than in Malaysia. It grows especially near the seashore. The leaves are simple.

PREPARATION   The pods contain a single large seed which is an excellent food, when boiled or roasted, even better than the chestnut.

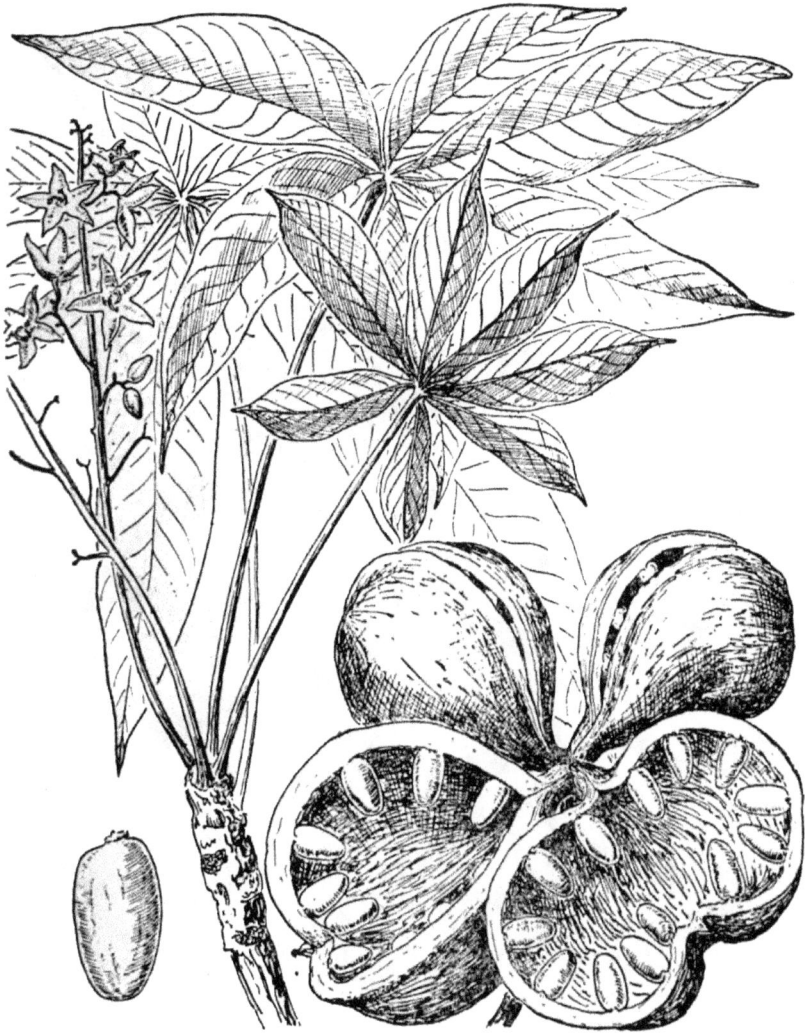

Poon tree, Wild almond, Kalumpang

Sterculia foetida L.   MALVACEAE

DESCRIPTION   This is a large tree with large red fruits. The numerous, nearly black seeds are rich in oil, the flavor somewhat suggesting the beechnut. Many other species of this genus occur in the forests but in most of them the leaves are simple. The seeds of all species are edible.

PREPARATION   The seeds may be eaten raw or roasted.

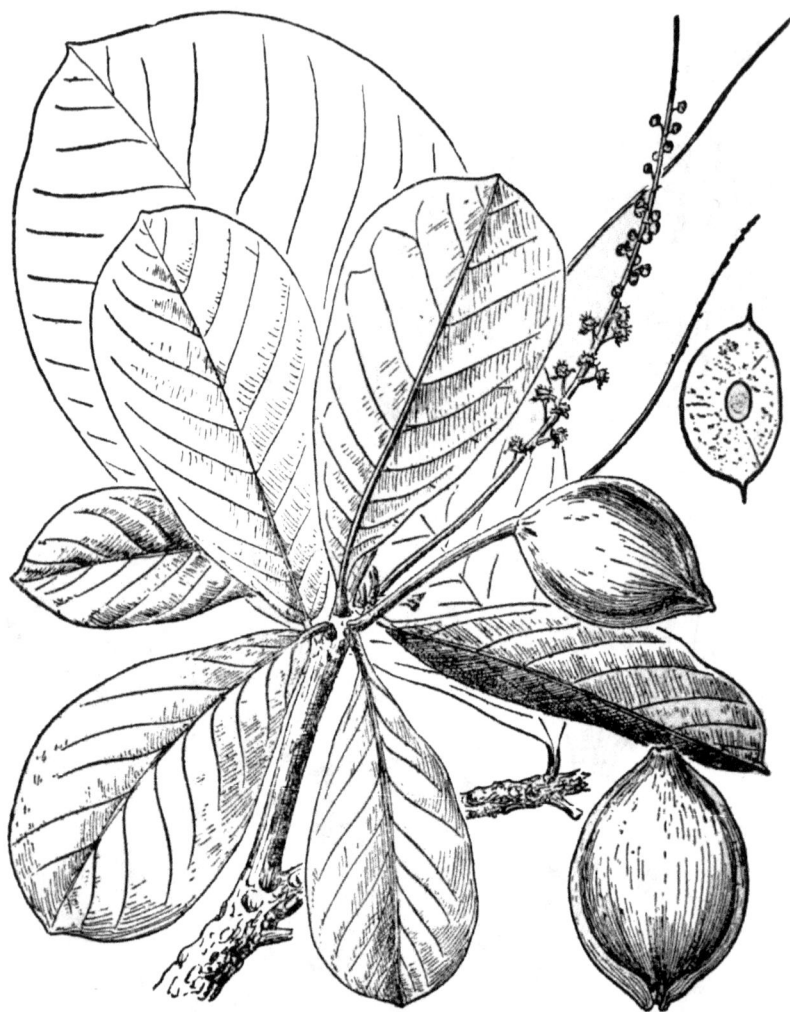

Indian almond

Terminalia catappa L.   COMBRETACEAE

DESCRIPTION   This is a large tree, generally found along the seashore but
                    sometimes planted inland. Not uncommonly planted as a
                    shade tree. Some of the leaves are usually red.

PREPARATION   The fruits contain a single fair-sized seed which is of
                    excellent flavor and may be eaten in any quantity. Many
                    other species of *Terminalia* occur in the forests, and the
                    seeds of all are edible.

Candle nut, Lumbang

Aleurites moluccanus (L.) Willd.　　　　　EUPHORBIACEAE

DESCRIPTION　　This is a common tree which may be identified by the foliage which is often pale green in contrast to that of other trees, or by the small greenish white flowers.

PREPARATION　　The fruits contain a single, hard-shelled seed rich in oil, and which may be eaten after roasting.

Queen sago

Cycas circinalis L.   CYCADACEAE

DESCRIPTION	This is a palm-like plant with a rather rough stem and very stiff leaves, sometimes found in the forest, more often near the seashore.
PREPARATION	The very young leaves (B), which are seasonal, may be eaten cooked as asparagus. The trunk yields a kind of sago but its extraction is difficult.
WARNING	The large seeds are used as food in times of scarcity but they are very poisonous unless properly prepared. The flesh is crushed or grated and soaked in water, with frequent changes of water, it being reported that this process should cover several days. The soaked material may be made into cakes and baked. Whenever possible, consult informed natives if the seed is to be used as food.

Kanari, Pili

Canarium indicum L.   BURSERACEAE

SYNONYM	*Canarium commune* L.
DESCRIPTION	There are many kinds of *Canarium* in Malaya, the genus extending eastward to Fiji and Samoa. They occur in the forests and are mostly of large size. The sticky resinous sap of the bark is distinctly fragrant or aromatic. The very hard inner parts of the fruit are usually more or less triangular in cross section, often pointed at the ends.
PREPARATION	The single, often fairly large, oily seed is well-flavored, and may be eaten raw or roasted.

Lotus and Water lily

Nelumbo nucifera Gaertn. (A)   NYMPHAEACEAE

Nymphaea spp. (B)

DESCRIPTION	These plants both occur only in shallow freshwater lakes and in slow streams. The lotus flowers are pink and the water lily flowers vary from white to pink or pale blue.
PREPARATION	The large seeds of the lotus (A) are excellent when boiled or roasted, while the large roots, found in the mud, may be cooked and eaten.

Pigeon pea

Cajanus cajan (L.) Millsp.   FABACEAE

DESCRIPTION   This is a small shrub or shrublike plant, 5 or 6 feet high, which is sometimes cultivated, but more often wild and frequently rather abundant. It occurs in open places, never in the forests. The flowers are yellow.

PREPARATION   The beans are edible but must be thoroughly cooked.

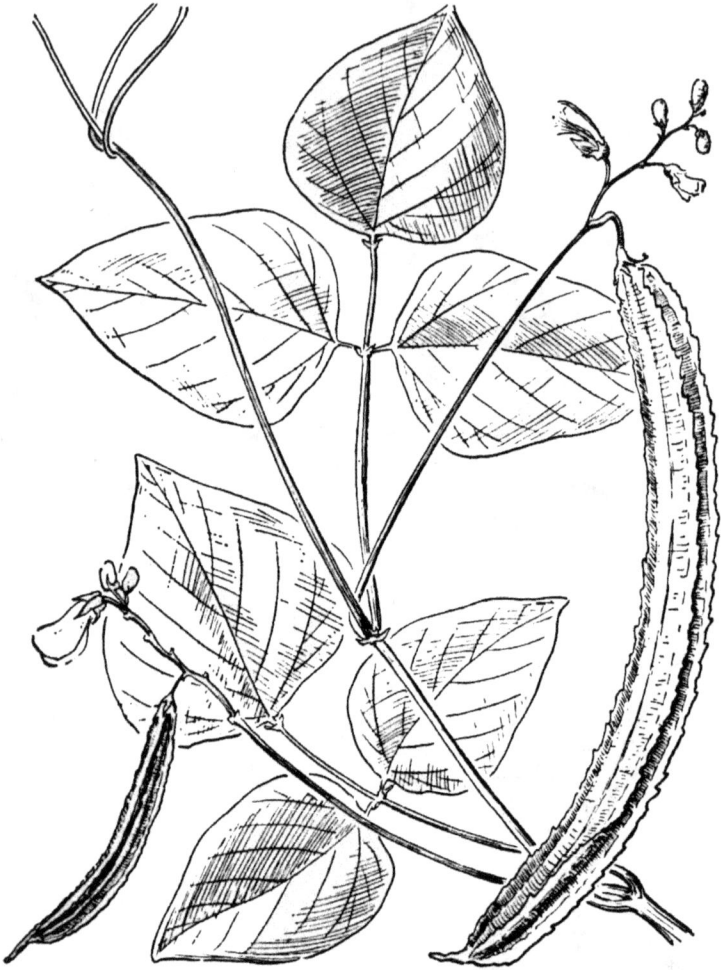

Asparagus bean

Psophocarpus tetragonolobus (L.) DC.   FABACEAE

DESCRIPTION   This is a twining vine from somewhat tuberous roots, the flowers are fairly large, and light violet-blue in color. The pods are 6 to 10 inches long, with four rather thin longitudinal wings. This is chiefly planted, but is sometimes found in fencerows.

PREPARATION   The tender pods, cooked as one would string beans, are an excellent vegetable. The mature seeds may be roasted and eaten.

Hyacinth bean

Lablab purpureus (L.) Sweet   FABACEAE

SYNONYM	*Dolichos lablab* L.
DESCRIPTION	This vine bears violet or white flowers. The young pods are often somewhat pink or reddish. The seeds are white, yellowish with black dots, or black with white dots. It is often cultivated, and frequently found wild in thickets.
PREPARATION	The young pods, an excellent vegetable, may be cooked and eaten as one would string beans, and even the flowers and young leaves may be cooked and eaten, as well as the ripe seeds.

Lima bean

Phaseolus lunatus L.   FABACEAE

DESCRIPTION	This vine has small flowers, greenish, sometimes white or violet within. It cultivated and naturalized, the wild form occurring in thickets.
PREPARATION	The very young tender pods may be cooked an eaten as one would prepare string beans.
WARNING	The mature seeds are often very poisonous (hydrocyanic acid)and one must be very careful when dealing with the wild forms, especially those with black seeds. The seeds vary in size and in color, ranging from white to brown or mottled or even jet black. The mature seeds of these wild forms may be eaten only after greatly prolonged cooking with many changes of water.

Peanut, Mani

Arachis hypogaea L.   FABACEAE

DESCRIPTION   The common peanut is often planted, especially in sandy soils.

PREPARATION   The fruits, borne under the surface of the ground, are very nutritious and the seeds may be eaten raw or cooked.

Physic nut (*Jatropa curcas*)

Poisonous Plants

Castorbean (*Ricinus communis*)

Physic nut

Jatropha curcas L. **EUPHORBIACEAE**

DESCRIPTION	This is a very common shrub found in hedges and fencerows.
WARNING	The rather large seeds are poisonous and violently purgative, not to be eaten under any circumstances.

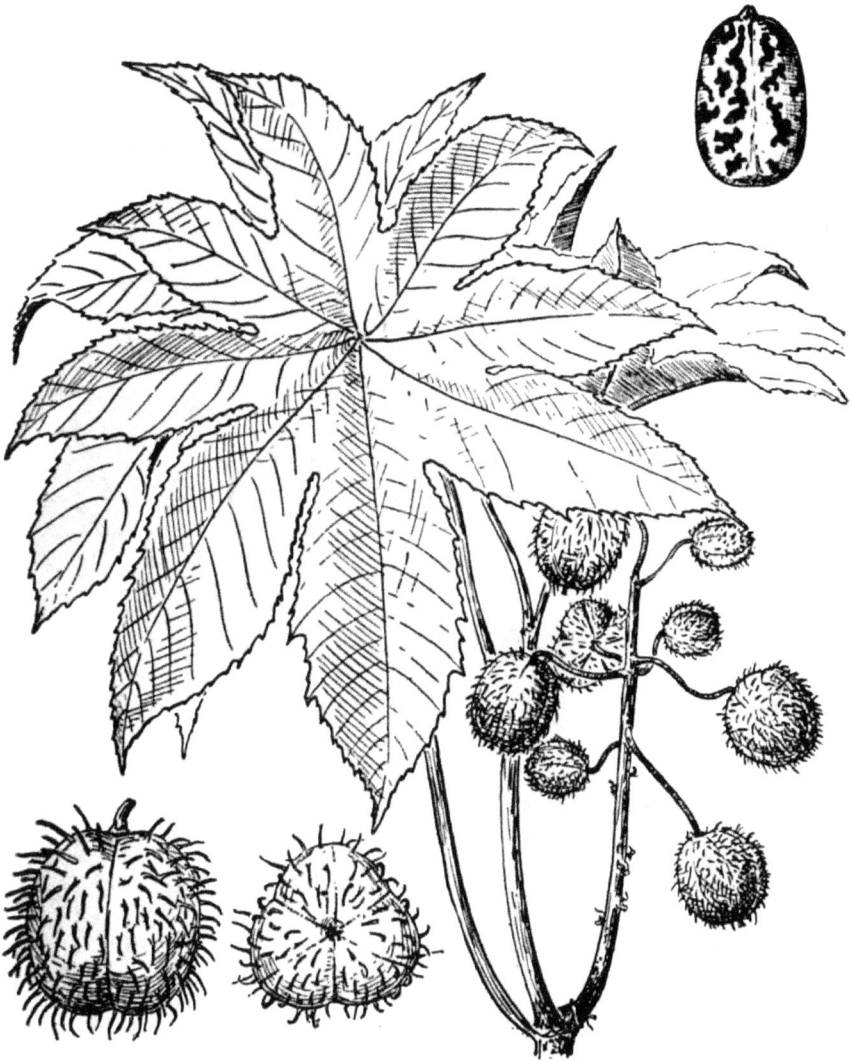

Castorbean, Castor oil plant

Ricinus communis L.   EUPHORBIACEAE

DESCRIPTION   This is a common, coarse, erect shrub or shrublike plant with large totea eaves. It is found in thickets and open places.

WARNING   The seeds are poisonous and a violent purgative, not to be eaten under any circumstances.

Tree nettle

Laportea spp.

URTICACEAE

DESCRIPTION   These shrubs, or small trees, grow in secondary forests and thickets, some species in the high forest. There are many species. The leaf-edges, nerves, leaf-stalk, flower- and fruit-bearing parts are supplied with stiff, very sharp, stinging hairs, often not conspicuous. These stinging hairs (A), seated on thin bulbs, are filled with intensely irritating sap. On light contact with the skin one immediately has the sensation of having touched a very hot iron, due, apparently, to the formic acid in the hair-sap. While intensely painful, the sting is normally not dangerous.

Tree nettle

Laportea spp.

URTICACEAE

DESCRIPTION

This plant is one of the forms with larger, broader leaves from the southwestern Pacific area. The stinging properties and local names are the same as for the narrower-leafed form (c above). The numerous small flowers are greenish or greenish white in all these tree nettles. (A) shows a stinging hair enlarged. The curious thing about these tree nettles is that if one grasps the leaves very firmly the result is usually little or no stinging, the stinging hairs being crushed. A light touch, however, usually results in an immediate burning sensation. Some species are much more irritating than others.

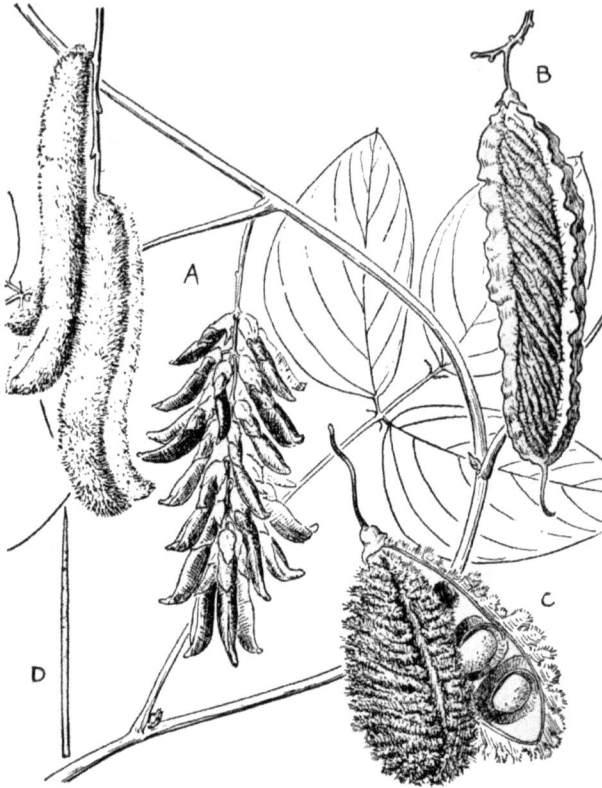

Cowhage

Mucuna pruriens (L.) DC.　　　　　　　　　　**FABACEAE**
Mucuna biplicata Kurz
Mucuna mollissima Kurz

DESCRIPTION　　　These vines occur in thickets and secondary forests, usually not in the high forest. The flowers are greenish white to very dark purple or even red. A number of different species occur in Malaya, some without stinging hairs. Parts of the flowers and the pods are covered with many stiff, easily detached, stinging hairs (C, much enlarged). While distinctly irritating, they are not poisonous, being mechanical irritants.

WARNING　　　One should avoid getting these stinging hairs into the eyes, as then they cause intense inflammation and may be very dangerous.

Marking-nut tree

Semecarpus spp.   ANACARDIACEAE

DESCRIPTION   These shrubs, or sometimes small trees, grow to a height of 30 feet and chiefly occur in thickets and in the secondary forests; many different kinds being known. Some of the species are reputed to cause bad skin eruptions on contact, or from the sap if the trees are cut down. The fleshy swollen basal parts of the fruits are usually dark purple and can be eaten with safety. The plant belongs in the same family with poison ivy and poison oak.

WARNING   Treatment, if one is poisoned, is the same as that indicated for poison ivy infections. These plants are not very dangerous.

Flowers, Cowhage (*Mucuna pruriens*)

Plants Used to Stupefy Fish

Sea poison (*Barringtonia asiatica*)

PLANTS USED TO STUPEFY FISH

In many parts of the region covered by this manual portions of several different kinds of plants are used to stupefy fish, both those found in tide pools and in small streams. The methods vary, but the usual one is to pound or crush the plant parts used, mix| with water, and throw a sufficiently large quantity of the material into pools which the fish inhabit. In streams, the material is always placed at the upper end of a quiet pool, thus permitting the current to spread it. Usually large quantities of this mixture must be used. The fish are suffocated, and usually come to the surface belly up; they can then be taken easily. The material used for this purpose in no manner affect the flesh, and fish thus secured can be eaten safely.

The most commonly used and most commonly available plants are different types of *Derris*, all woody vines, widely known as *tuba*. The fish poison is most abundant in the roots, but in preparing the material both the roots and other parts of the plants are pounded and thrown into the water. *Derris* is the most efficient of the various plants used.

The large one-seeded fruits of *Barringtonia* asiatica are also used. This is a large tree with large, very smooth leaves, pink flowers, and one-seeded fruits that are square in cross-section. Its natural habitat is only along the seashore of Malaysia and parts of Polynesia. The large solitary seed is mashed and thrown into the pools where the fish occur.

Another often used plant is the shrub or small tree *Croton tiglium*. The small seeds are crushed and used as in *Barringtonia*. The species is commonly found about settlements, near houses, and is naturalized in waste places; it does not occur in the forests.

In Polynesia and Micronesia the most used fish poison is *Tephrosia purpurea*, a small shrub or somewhat woody herb with small purple flowers and a small flat pod. The whole plant is pounded or crushed and thrown into the water.

The word *tuba* which is widely used in Malaya and applied to a number of totally different plants, usually indicates a plant that may be used for stupefying fish.

Croton oil plant

Croton tiglium L.   EUPHORBIACEAE

DESCRIPTION	This shrub, or small tree, is cultivated and spontaneous. The seeds are used chiefly for poisoning fish.
WARNING	A very violent purgative. Not to be eaten under any circumstances.

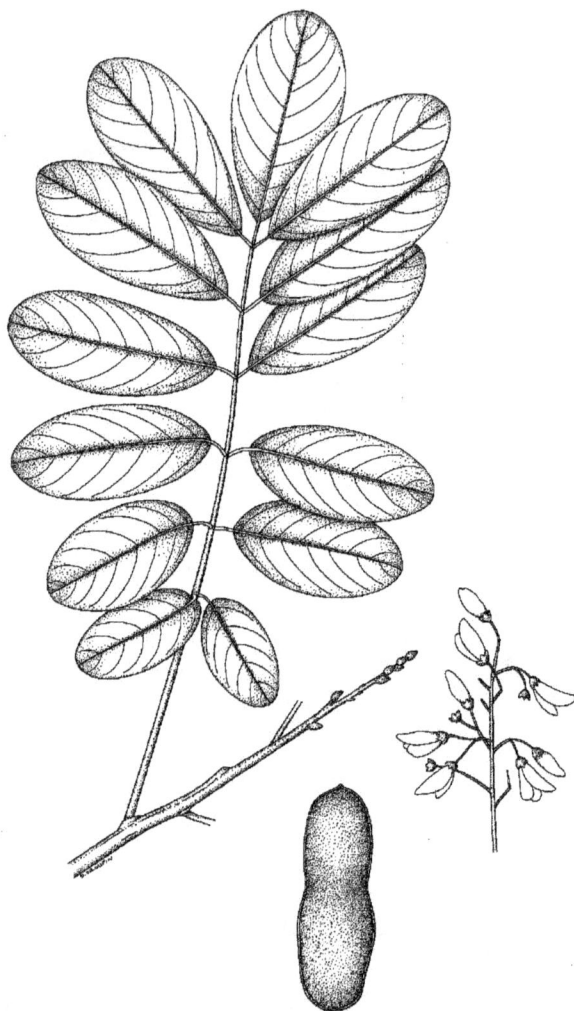

Derris

Derris elliptica (Wall.) Benth.   FABACEAE

DESCRIPTION These are all woody vines, with flowers resembling those of the bean, and narrowly winged pods, occurring chiefly in thickets and secondary forests. This is a most efficient fish poison; its use for suffocating fish in slow streams, pools, and even tidal pools is widely known. There are many different species of the genus, some more potent than others. The parts used are chiefly the crushed roots, but sometimes the crushed branches and leaves are also used.

Wild indigo

Tephrosia purpurea (L.) Pers.   FABACEAE

DESCRIPTION   This is a small, branched shrub or shrubby herb, growing in open places. The small flowers are purple. It is widely used as a fish poison, especially in Polynesia. The whole plant is crushed and thrown into the pools where fish occur.

Sea poison

Barringtonia asiatica (L.) Kurz   LECYTHIDACEAE

DESCRIPTION These large trees grow on the seashore. They bear large smooth leaves, large pink flowers with many stamens, and large fruits which are square in cross-section and which contain a single large seed. The crushed seeds are used for killing fish in pools.

Index

Acalypha
    indica, 68
    wilkesiana, 69
Acrostichum aureum, 6
Agati, 72
Agave cantala, 77, 78
Aleurites moluccanus, 117
Alocasia macrorrhiza, 11
Alternanthera sessilis, 53
Alugbati, 54
Amaranth, 51
Amaranthus, 51
Amischotolype, 50
Amischotolype mollissima, 50
Amorphophallus paeoniifolius, 13
Ampalaya, 48
Anacardium occidentale, 90
Annona
    muricata, 92
    reticulata, 93
    squamosa, 91
Antidesma bunius, 107
Arachis hypogaea, 125
Arenga, 22
Arrowroot, 41
Artocarpus
    altilis, 83
    heterophyllus, 84
    integer, 85
    rigidus, 86
Asparagus bean, 122
Averrhoa
    bilimbi, 102
    carambola, 102

Balsam vine, 48
Bamboo, 31
Banana, 81
Barringtonia asiatica, 135, 140
Basella alba, 45, 54
Bayabas, 89
Bignai, 107

Bilimbi, 102
Black nightshade, 59
Boerhaavia diffusa, 58
Breadfruit, 83
Buck yam, 39
Bulb yam, 37
Buri palm, 22

Cajanus cajan, 121
Calamus, 21
Canarium indicum, 119
Candle nut, 117
Cantala, 77, 78
Carambola, 102
Carica papaya, 82
Caryota, 22
Cashew, 90
Cassava, 35
Castor oil plant, 127, 129
Castorbean, 127, 129
Celosia argentea, 52
Ceylon spinach, 45, 54
Champedak, 85
Cheese fruit, 76
Chico, 95
Climbing fern, 6
Cock's comb, 52
Coconut, 15, 24
Cocos nucifera, 15
Coix lacryma-jobi, 29, 29
Colocasia esculenta, 7, 9
Commelina, 49
Copperleaf, 69
Coral tree, 71
Corypha, 22
Cowhage, 132
Croton oil plant, 137
Croton tiglium, 137
Custard apple, 93
Cyanotis, 49
Cyathea, 4
Cycas circinalis, 118

Cynometra cauliflora, 104
Cyrtosperma merkusii, 12

Daeking tree, 106
Dayflower, 49
Derris, 138
Derris elliptica, 138
Dioscorea
    alata, 36
    bulbifera, 37
    esculenta, 38
    hispida, 40
    pentaphylla, 39
Diplazium esculentum, 1, 5
Drop tongue, 10
Duck-lettuce, 61
Duhat, 96
Dwarf copperleaf, 53

Eichhornia crassipes, 63
Eleocharis dulcis, 44
Emilia sonchifolia, 64
Erechtites, 65
Erythrina variegata, 71

False pickerelweed, 62
Fireweed, 65
Fishtail palm, 22

Gnemon tree, 106
Gnetum gnemon, 106
Goa yam, 38
Golden leather fern, 6
Greater yam, 36
Ground cherry, 110
Guava, 89
Guyabano, 92

Hairless clearweed, 55
Horseradish tree, 70
Hummingbird tree, 72
Hyacinth bean, 123

Indian acalypha, 68
Indian almond, 116
Indian camphorweed, 67
Indian mulberry, 76
Inocarpus fagifer, 114

Ipomoea
    aquatica, 60
    batatas, 34

Jack fruit, 84
Jambolan, 96
Jatropa curcas, 126, 128
Job's tears, 29

Kalumpang, 115
Kamansi, 83
Kanari, 119
Kapayas, 82

Lablab purpureus, 123
Langka, 84
Lansium parasiticum, 79, 88
Lanzone, 88
Laportea, 130, 131
Lau lau, 97
Lilac tasselflower, 64
Lima bean, 124
Lotus, 120
Luffa, 47
    acutangula, 47
    cylindrica, 47
Lumbang, 117
Lycopersicum esculentum, 109

Malay apple, 98
Malunggai, 70
Mangifera indica, 94
Mango, 94
Mani, 125
Manihot esculenta, 33, 35
Manilkara zapota, 95
Manioc, 35
Maranta arundinacea, 41
Marking-nut tree, 133
Melinjo, 106
Metroxylon, 18
Momordica charantia, 48
Monkey jackfruit, 86
Monochoria
    hastata, 62
    vaginalis, 62
Morinda citrifolia, 76
Moringa, 70

Moringa oleifera, 70
Mucuna
  biplicata, 132
  mollissima, 132
  pruriens, 132, 134
Musa, 81

Nam-nam, 104
Nanka, 84
Nelumbo nucifera, 111, 120
Nephelium
  lappaceum, 87
  mutabile, 87
Nipa palm, 25
Noni, 76
Nymphaea, 120
Nypa fruticans, 25

Ottelia alismoides, 61

Pachyrhizus erosus, 42
Paco, 5
Palmgrass, 30
Pandan, 105
Pandanus tectorius, 105
Pangi, 112, 113
Pangium edule, 112, 113
Papaw, 82
Papaya, 82
Paracress, 66
Peanut, 125
Pemphis, 74
Pemphis acidula, 74
Phaseolus lunatus, 124
Physalis, 110
Physic nut, 126, 128
Pigeon pea, 121
Pilea glaberrima, 55
Pili, 119
Pluchea indica, 67
Polynesian arrowroot, 43
Polynesian chestnut, 114
Polynesian plum, 101
Poon tree, 115
Portia tree, 73
Portulaca oleracea
Psidium guajava, 89
Psophocarpus tetragonolobus, 122

Pulusan, 87
Punarnava, 58
Purslane, 56

Queen sago, 118

Rambutan, 87
Rattan palm, 21
Red spiderling, 58
Ricinus communis, 127, 129
Rose apple, 99

Saging, 81
Sago palm, 18
Salacca zalacca, 20
Salak, 20
Sampalok, 103
Sandoricum koetjape, 100
Santol, 100
Sapodilla, 95
Schismatoglottis calyptrata, 10
Screw pine, 105
Sea lemon, 108
Sea poison, 140
Seaside purslane, 57
Semecarpus, 133
Sesbania grandiflora, 72
Sesuvium portulacastrum, 57
Setaria palmifolia, 30
Solanum americanum, 59
Sour sop, 92
Spilanthes acmella, 66
Spondias dulcis, 101
Stenochlaena palustris, 6
Sterculia foetida, 115
Sugar palm, 22
Swamp fern, 4
Swamp morning-glory, 60
Sweet potato, 34
Sweet sop, 91
Syzygium
  aqueum, 97
  cumini, 96
  jambos, 99
  malaccense, 98

Tacca leontopetaloides, 43
Tallow wood, 108

Tamarind, 103
Tamarindus indica, 103
Tapioca, 35
Taro, 7, 9
Tephrosia purpurea, 139
Terminalia catappa, 116
Thespesia populnea, 73
Toothache plant, 66
Tournefortia argentea, 75
Tree ferns, 3
Tree heliotrope, 75
Tree nettle, 130, 131

Water cherry, 97
Water chestnut, 44
Water hyacinth, 63
Water lily, 120
Wild almond, 115
Wild indigo, 139
Wild tomato, 109
Wild yam, 40

Ximenia americana, 108

Yam bean, 42
Yellow plum, 108